AESOP'S ANIMALS

AESOP'S ANIMALS

THE SCIENCE BEHIND THE FABLES

Jo Wimpenny

BLOOMSBURY SIGMA
LONDON · OXFORD · NEW YORK · NEW DELHI · SYDNEY

BLOOMSBURY SIGMA
Bloomsbury Publishing Plc
50 Bedford Square, London, WC1B 3DP, UK
29 Earlsfort Terrace, Dublin 2, Ireland

BLOOMSBURY, BLOOMSBURY SIGMA and the Bloomsbury Sigma logo
are trademarks of Bloomsbury Publishing Plc

First published in the United Kingdom in 2021

A catalogue record for this book is available from the British Library

Library of Congress Cataloguing-in-Publication data has been applied for

ISBN: HB: 978-1-4729-6691-9; eBook: 978-1-4729-6693-3

2 4 6 8 10 9 7 5 3 1

Typeset by Deanta Global Publishing Services, Chennai, India
Printed and bound in Great Britain by CPI Group (UK) Ltd,
Croydon CR0 4YY

Illustrations by Hana Ayoob

Bloomsbury Sigma, Book Sixty-eight

To find out more about our authors and books visit www.bloomsbury.com
and sign up for our newsletters

For Mum and in memory of Dad – you always believed in me, and you taught me to never give up. Thank you for everything.

Contents

Preface

'Just one more,' the little girl pleads. 'Pleeeeeaase.'

Her brother joins in, though he's sleepy. 'Yes, please, Daddy, one more.'

He sighs, reminds them that's what they said last time, and the time before, and they giggle conspiratorially.

'Alright, little monkeys, but this is the last one. And then lights out – it's sleep time!' He picks up the book and it falls open at a well-worn page. 'Ahh, here we go, one of my favourites: 'The Fox and the Crow'.' He shows the pages to his children, who gaze at the lavish illustrations, and begins.

'A Crow having stolen a bit of meat, perched in a tree and held it in her beak. A Fox, seeing this, longed to possess the meat himself ...'

'Daddy, crows are clever, aren't they?' It's his son, but before he can respond his daughter chips in too.

'No, but actually foxes are cleverer than crows. Foxes are the cleverest, aren't they?'

He smiles and continues. They snuggle down into their duvets – foxes, crows and talking beasts filling their minds as they drift to sleep.

★ ★ ★

Aesop's fables are the stories of our youth. How many of us were these children, revelling in a version of the fables at bedtime? I'd wager quite a few, even if you don't remember the details. And I bet you're familiar with sayings like 'slow and steady wins the race', 'the lion's share', 'sour grapes' or 'crying wolf'. They all have their roots in Aesop's fables, a collection of short stories penned some 2,500 years ago.

It's good to manage expectations early, so I want to be clear that this isn't a book about Aesop. If you're hoping to learn more about the man behind the fables, this probably isn't the book for you. If, on the other hand, you have idly wondered whether foxes or crows are cleverer, if wolves really are deceptive or a tortoise could ever actually beat a hare in a race, then read on!

Some introduction to Aesop is needed, of course. The popular view is that he was a deformed, ugly slave who lived between 620–564 BC and won his freedom by telling stories to the royal courts. Some prominent Greek figures, including Aristotle and Herodotus, referred to Aesop in their works, suggesting that they at least believed he was a real person. Many scholars, on the other hand, have pointed out the inconsistencies in different accounts of Aesop's life, suggesting instead that he was a fictional character, created soon after the collection of stories known as Aesop's fables rose to popularity in ancient Greece. Throughout this book I write as if he was real, because it's easier than needing to caveat every mention. But really, and this might sound bizarre for a book entitled *Aesop's Animals*, whether Aesop was a real person is not of critical importance. What matters is the reality of Aesop's fables, the fact that, no matter how they came to life, this collection of ancient stories has stood the test of time.

Although today, Aesop's fables are largely marketed for parents to read to their children, they were originally created as a form of social commentary, a tongue-in-cheek way to highlight human foibles and provide moral instruction. They were first printed in English in the fifteenth century and a further push came in the late seventeenth century, when philosopher John Locke published *Some Thoughts Concerning Education*, in which he mused about books that could help children to learn without filling their heads with 'useless trumpery'. Locke concluded: 'I think Aesop's Fables the best, which being Stories apt to delight and entertain a Child, may yet afford useful Reflections to a grown Man.' Their popularity soared and by the eighteenth century, publisher Robert Dodsley commented that: 'Along with the Bible and *The Pilgrim's Progress*, Aesop may be said to have occupied a place on the meagre bookshelf of almost every cottage.' Today, multiple copies are likely to be found in the children's section of any bookshop. Our world has changed beyond recognition and yet, remarkably, Aesop's fables still have an important place in it.

★ ★ ★

As a zoologist, animal behaviour has always fascinated me. The fact that you can study it scientifically, on the other hand, eluded me until an A level Biology lesson in which my teacher (aptly called Mr Bird) described how robins and male sticklebacks respond aggressively to the colour red. I remember being enthralled to learn that people had studied these animals displaying towards postal vans and taxidermy robins; it was as if a switch had been turned on in my brain.

Today, the science of animal behaviour is flourishing, meaning that our knowledge is constantly advancing. Some of it gets reported, but much of it doesn't, and the confines of a newspaper article make it difficult to provide much in the way of wider or historical context. My overarching aim for this book is to bring to life some of the most remarkable discoveries about what animals do and why, in the form of stories that interweave science with history, expert perspectives and plenty of fascinating facts. In doing so, my deepest wish is that you will share some of that same sense of wonder.

So why am I, a scientist, bringing in the apparent fiction of Aesop's fables – why not just pull together some of the most incredible behavioural discoveries? Initially, my use of the fables was simply a storytelling aid – a popular hook to help build a bridge to the science. That's still true, but over time I've recognised a deeper link with Aesop's fables, which comes down to their remarkable longevity. It's impossible to sum up how much the world has changed in the past 2,500 years; how different twenty-first-century life looks to the lives of the ancient Greeks or indeed any ancient civilisation. And yet Aesop's fables form a thread of continuity among the maelstrom of history. They have well and truly infused our collective consciousness, as preserved and also subsumed into other folk stories around the world as they are. It's an incredible testament to their popularity, yet the bizarre consequence is that we continue to tell stories that were never created for modern times. I don't mean the morals: while dated, they're not exactly controversial lessons for the present day, but our knowledge and understanding of animals today is worlds away from Aesop's time. This matters because these stories, which engage us from such an early age, naturally also shape our emerging knowledge about animal natures.

Aesop's fables are wholly, unashamedly anthropomorphic, meaning that the animals are all bestowed with human qualities and emotions. That was their very purpose. Aesop was not an animal behaviour scientist (or, indeed a scientist at all, given that the discipline had not yet emerged) and his fables weren't intended to reveal anything about animal minds. Animal characters are often used because they can provide useful props for stories: they offer the perfect balance of familiarity and otherness that can enable potentially sensitive topics or issues to be communicated in a way that is simultaneously detached. But there was also a pattern to Aesop's use of different animals for different roles – he, together with other fabulists, was capitalising on existing beliefs about animal natures, such as those of the ancient Egyptians, where animals were venerated and worshipped. It's impossible to pin down when humans started attributing certain behavioural traits to animals, but the roots surely go deeper than this. For example, a look at the 35,000-year-old art of the Chauvet caves in southern France demonstrates that hunter-gatherer societies were highly skilled observers of animals. They needed to be to live alongside them, to predict their behaviour so that they could more effectively hunt them or avoid being hunted. The lack of a scientific framework in which to interpret animal behaviour certainly doesn't mean a lack of observational or cognitive skills to think about what animals were doing and why.

We've continued to propagate the same character traits as Aesop and those before him – the crafty fox, deceptive wolf, stupid donkey. It's handy really, when consistent animal characters are used in predictable ways, because they fit with the reader's preconceptions; as a result, stories can be kept short because character traits don't need to be set up. By using a fox in a fable about thinking on one's feet, for example, the story can be much shorter than if the character was a fictional human who needed to first be described before he did whatever the fable was trying to convey.

The problem is that our folkloric knowledge doesn't always match up with reality. Which doesn't matter for unicorns, dragons or other mythical creatures. And it shouldn't be a problem for animal stories either, until fact and fiction start to get muddled. Some animals have become so entwined with their fictional, human-like characters that for many people they've become one and the same. As a result, the fables, which surely played a role over the centuries in forming these ideas, became accepted as truth. In Dodsley's words:

> The truth is, when moral actions are with judgement attributed to the brute creation, we scarce perceive that nature is at all violated by the fabulist. He appears, at most, to have only translated their language. His lions, wolves, and foxes, behave and argue as those creatures would, had they originally been endowed with the human faculties of speech and reason.

Most scientists agree that we're living through the sixth mass extinction event and that it's being caused by our activity on the planet. Aesop would not have been able to imagine it, but animals need our help today more than ever before. And we can't do that unless we put folkloric beliefs to one side. At a time when so many animals are threatened, when we've already stacked the odds against them by destroying their habitats, persecuting and hunting them, how unhelpful it is that we stick derogatory labels on some of them too, accusing them of character flaws that were based on human faults all along.

For these reasons, I think it's important that we consider our inbuilt preconceptions of animal natures. Because we do hold them, all of us, and without a doubt Aesop's fables and other stories played a role in their genesis. Nonetheless, while the seeds may have been sown during childhood, our beliefs are not fixed; they can be adjusted and even reformed as a result of life experiences. Of course, when you hear something that supports an existing belief, it will be strengthened. Our brains can make connections before we have time to think about them, mental shortcuts called 'heuristics' that help to free up cognitive capacity for other things. It's an essential process that prevents information overload, helping us to make sense of the world. But it can also be a hindrance because those mental shortcuts are not always spot on. Sometimes they lead to systematic errors of judgement, called cognitive biases, which predispose us to behave in certain ways.

Many cognitive biases have been suggested and it's good to be aware of their existence. Here though, I just want to focus on one, the confirmation bias. This is the tendency we all have to pay greater attention to things that fit with what we already believe. As open-minded as we think we are, our brains still zone in on things that reinforce existing beliefs and disregard what doesn't fit. We are 'cognitively lazy'. Evaluating all the evidence all the time takes a lot of conscious effort. It's because of the confirmation bias that we find it so easy to jump to conclusions about animal characters. One of my aims with this book is to create a little more friction – to challenge us all to pause next time we see a viral video of a 'clever crow' and think about what might actually be going on. Because the reality is that the science of animal behaviour is both more complex and more fascinating than these videos can reveal.

★ ★ ★

'The Crow and the Pitcher' fable was the original inspiration for this book, so we start there. As a zoologist who studied crow problem-solving for my DPhil, it's also the fable closest to my heart. Through that discussion and the eight chapters that follow, we'll explore different exciting topics in animal behaviour – tool use, future planning, self-recognition, imitation, deception and cooperation – as well as ask whether science supports our preconceptions of donkeys and foxes, and consider the reality of the tortoise and the hare. Each chapter closes with a fact-or-fiction verdict on whether the fable has been appropriately cast and, if not, which creature from the animal kingdom would be better placed to step into the role.

While science is at the core of this book, these are not comprehensive academic reviews. They are stories about science and while I hope that they are of interest to those working in the field, summarising every important study, person and theoretical advance would be a gargantuan effort for the most seasoned researcher. I see this as a wonderful testament to the status of animal behaviour in modern science: it's a huge and ever-expanding field.

On to the fables. I have used the version translated by George Fyler Townsend, which was first published in 1867. There are many others, and while there are some language differences between versions the essence of each fable and the animal character does not really change. I have, of course, only been able to include a small fraction of Aesop's fables in this book, so I can only apologise if your favourite is missing. However, when you next revisit this fable, think about how its animal characters are portrayed and if this matches your own modern knowledge about their natures.

Lastly, many animals are missing, some of which might be surprising for a book about animal behaviour. There is no dedicated chapter about whales, dolphins, elephants, octopuses, parrots or bees, all of which (and more) are currently revealing fascinating cognitive abilities. But there must be a limit, and I know there will be other opportunities to tell their stories.

I hope that you enjoy reading this as much as I've enjoyed writing it. See you on the other side.

He sat among the woods; he heard
The sylvan merriment; he saw
The pranks of butterfly and bird,
The humours of the ape, the daw.

And in the lion or the frog –
In all the life of moor and fen,
In ass and peacock, stork and log,
He read similitudes of men.

'Aesop', Andrew Lang (1844–1912)

CHAPTER ONE

The Crow and the Pitcher

A Crow perishing with thirst saw a pitcher, and hoping to find water, flew to it with delight. When he reached it, he discovered to his grief that it contained so little water that he could not possibly get at it. He tried everything he could think of to reach the water, but all his efforts were in vain. At last he collected as many stones as he could carry and dropped them one by one with his beak into the pitcher, until he brought the water within his reach and thus saved his life.

It's 29 July 2008 and something remarkable is about to happen inside a small room in a pretty, sleepy little Cambridgeshire village. Cambridge University researchers Chris Bird and Nathan Emery have set a puzzle for a male rook named Cook;

now they sit back and wait. Cook is presented with a 15cm-high clear Perspex tube attached to a board. It is half full of water and a juicy wax moth larva bobs at the surface. Cook peers into the tube but cannot stretch his head in far enough to reach the larva, which is one of his favourite treats. After inspecting the tube from the side, Cook picks up one of several stones that are scattered on the table. Holding it in his long grey bill, he takes it into the tube and drops it. He repeats the behaviour, again and again, until the larva is within reach. Finally, he claims his prize! For Cook the rook, this was just another day where he ate a tasty treat; for Bird and Emery, it was a lot more exciting. On his first attempt, Cook had replicated Aesop's fable.

More than two and a half millennia have passed between Aesop and Cook, and in that time the world has changed immeasurably. Our ideas and beliefs about our place in the world have seismically shifted, never mind our understanding of our fellow animals. Yet here was a thread, clear and true, spun by an ancient Greek storyteller and woven into the fabric of modern science.

Bird and Emery's study prompted many reactions. The media loved it and it was covered worldwide. Many researchers were excited by the demonstration of flexible problem-solving in rooks, stimulating follow-on studies in other species to examine the behaviour more carefully. Others were exasperated by what they saw as a fad in animal cognition research, questioning the validity of using a fable to structure scientific investigation. I saw an intersection between science and society – a way to combine facts with fiction to bring the latest research to life. This, in essence, was my aim: to tell stories about stories, grounded in science – and I needed to start with the crows.

★ ★ ★

The crow family (scientifically the *Corvidae* or corvids) includes more than 120 species of crows, ravens, rooks, magpies, jays, jackdaws, choughs, nutcrackers and treepies. Perhaps surprisingly, they sit within the same evolutionary group as songbirds (the *Passeriformes*) and while they don't sing a song like a blackbird they do have a remarkably diverse vocal repertoire, including mimicry. Within the family, the genus *Corvus* contains the crows, ravens and rooks – a grouping of about 45 species that are characterised as medium-to-large birds with mainly black plumage. Outside of this group, some other corvids are far more colourful, particularly in exotic locations (the Sri Lanka blue magpie is spectacular, as is the green jay of southern and central America). Even closer to home, the Eurasian jay is a visual treat with its rich chestnut plumage, cerulean wing edges and dapper black moustache; and I find plenty of beauty in the sleek iridescence of a magpie's tail or the pale blue eyes of a jackdaw.

Corvids are an astoundingly successful group, found everywhere except for the poles and the southern part of South America. As a result, everyone knows something about at least one of them, as Kaeli Swift, postdoctoral researcher at the University of Washington, explains: 'One of the most fascinating things about these birds is that because almost everywhere people have settled there is a species of crow or raven that lives alongside them, they have wormed their way into the religions and cultural storytelling of people worldwide.'

One of the things she would like to overturn is the perception that crows are portrayed negatively all over the world. 'That's just not true everywhere. Crows can mean very different things to different people and I wish more people appreciated the great nuance and uniqueness of their symbology to people around the world.'

It's true that in the Western world, we seem to have homed in on the less endearing corvid traits. For one, crows and ravens have no qualms about feasting on human bodies – they learned a long time ago to show up at gallows and battlefields, patiently waiting until they could land and take advantage of an easy meal. As a result, and particularly in medieval Europe at the time of the plague, they earned an association with death that has stuck to the present. Crows became omens of darkness, messengers from the underworld, resulting in widespread fear and persecution. These beliefs even worked their way into our language: the fifteenth-century *Book of St Albans* was the first to term a group of crows as a 'murder' and a group of ravens as an 'unkindness'. The sinister side of the crow family is perhaps amplified by their perceived intelligence: being a deathly emissary is bad enough, but one that can use its brain seems considerably worse. In Norse legend, for example, the two ravens Huginn and Muninn (roughly translated as 'thought' and 'memory') served as the god Odin's eyes in the world of men, and were depicted as both intelligent and bloodthirsty. That glossy all-black plumage probably doesn't help either – if crows and ravens were as brightly coloured as parrots, would they be as feared?

On the other hand, in North America the mythological figure Raven features frequently in the spirit stories of indigenous people – he is a creator, shapeshifter and mischief-maker. For many First Nation tribes, the intelligence of corvids was their greatest feature and medicine men would call upon Raven to provide clarity to visions. To the ancient Egyptians, crows symbolised faithful love, because already at this time people could see that they were strongly bonded and monogamous. In Chinese mythology, crows represented power and were associated with the sun – each of the 10 suns had a crow spirit and some stories tell of crows whose duty it was to carry the sun across the sky. The national bird of Bhutan is a raven, symbolising one of the country's most powerful deities. Here, the king wears a crown topped by an embroidered raven head and it was once a crime punishable by death to kill this bird.

Legend has it that if the ravens at the Tower of London were to ever leave, the kingdom would fall, so during his seventeenth-century reign King Charles II decreed that there must always be six resident birds. When Tower Hill was bombed during the Second World War, Winston Churchill ordered the birds' immediate replacement, bringing in ravens from the Welsh hills and Scottish Highlands. Their haunting croaks have echoed around the tower ever since. Today, the Tower's seven ravens (there must always be one spare) are cared for full-time by the 'Ravenmaster'. The birds live for up to 20 years and each has its own character traits. Raven George, for example, developed a taste for eating television aerials in the mid-1980s and had to be dismissed for 'unsatisfactory behaviour'; Raven Grog was last seen outside the Rose and Punchbowl pub in 1981; and Merlina, who sadly disappeared in early 2021, was known as a bit of a diva and would occasionally play dead for tourists. In fact, research on the social and physical intelligence of ravens has revealed some remarkable abilities, which we'll look at later.

Crows may not have the visual wow factor of a bird of paradise, nor the delicate beauty of a hummingbird, but because they are common across the world, reports of their intellectual prowess cause us to look at them again, to recall and share our own stories about crow smarts. A 2012 home video of a crow 'snowboarding' down a roof in Russia became an instant social media hit: with millions of views, it continues to astound those who discover it. Other, equally fascinating videos of crow behaviour exist: the crows that repeatedly slide down snowy slopes or the rolling somersaults of ravens during display flights – far from being sinister omens of death, these birds appear to be having fun.

Aesop depicted the crow as inventing an insightful solution to the problem of obtaining water, later supported by first-century naturalist Pliny the Elder's report of a raven piling up stones in a memorial urn. But to really know whether Aesop's crow is based on fact or fiction we must delve into the science and examine the facts of corvid cognition. Luckily, the past three decades provide rich pickings.

★ ★ ★

Cambridge University is a leading centre for research on animal minds. Much of this stems from the work of Nicky Clayton, professor of comparative cognition, who began and continues to oversee the research. It was during her time as a researcher at the University of California in Davis that Clayton became seriously interested in corvids. In the autumn months, when new students hurried nervously around campus, Clayton – horrified that most of her colleagues chose to spend lunchtime at their desks – sat outside and ate in the sunny campus grounds. If you've spent time in California you probably will have noticed some large, screechy, blue-grey birds. These are western scrub jays and Clayton noticed that they were busily burying acorns. This in itself was unremarkable, but she was interested to see some of the birds returning to their stashes, digging them up and reburying them. She had questions: how did they remember where they had buried their food? How long could they remember for? And were they moving their stashes to prevent them from being stolen? Clayton moved to Cambridge University and founded a captive colony of western scrub jays to investigate all these questions and more (we'll return to this in Chapter 8). Subsequently, the group's research has diversified to include studies of several other corvid species, including Eurasian jays, jackdaws, rooks and New Caledonian crows, as well as non-corvids and human infants.

Bird was studying rook behaviour for his PhD at Cambridge, supervised by Emery. Rooks are a common British corvid and often confused with crows: they're similar sized but distinguished by their steeper foreheads, long grey bills and pale, featherless faces. If you've ever driven past a ploughed field and seen what looks like lots of crows pecking around, they were more likely rooks; the lack of face feathers is an adaptation for plunging their bills deep into the earth for grubs. They're also more social than crows, nesting communally in 'rookeries', which can number hundreds, if not thousands, of birds. Rooks are also quite common at motorway services, where they can make a nuisance of themselves by pulling rubbish bags out of bins. It was while researching historical records of crow problem-solving that Bird came across an anecdotal report of a rook plugging up a hole in its aviary and causing water to pool. It triggered a recollection of Aesop's fable, so he started to think about testing it scientifically.

The Cambridge rooks weren't short of water − instead their motivation took the form of a fat wax moth larva that floated at the water's surface. The required behaviour was nonetheless the same: the bird needed to raise the water level to bring the larva within beak reach. Bird was astounded that Cook the rook solved the problem on his first trial. He tested him again to see whether the behaviour was a fluke. It wasn't. Bird then tested Cook's mate, Fry − who was in the same aviary − and she also solved the problem on her first trial, but after five trials refused to participate further.[*] Intrigued, Bird and Emery tested a second pair, Connelly and Monroe,

---

[*] This sometimes happens in animal research. Bird had been struggling to find a reliable way of making the worms float and had resorted to using either the most bloated worms in the batch or dipping the worms in oil to increase their buoyancy. The problem is that to a rook, bloated worms may be bad worms, and oily worms are a bit weird. For whatever reason, Fry decided the worms on offer weren't valuable enough to carry on in Bird's study.

adapting the apparatus so the larva was tacked to a small piece of cork. 'It was quite tricky in some cases to get the apparatus quite right,' reflects Bird. 'One bird, for example, tried to get the worm so much without using the stones that it tipped itself upside down and got its head stuck in the tube.' Connelly and Monroe were initially stumped, but quickly got the hang of the task on their second trials and consistently thereafter. Not only could they solve the basic problem, they also made sensible choices on variations of it: preferring to drop stones into a tube full of water rather than sand and preferring to drop heavy rather than light objects.

Bird and Emery published their findings in a top journal, commenting that the results provided 'the first empirical evidence that a species of corvid is capable of the remarkable problem-solving ability described more than two thousand years ago by Aesop'. The media loved it and Bird found himself being interviewed on national television. Aesop's fables enjoyed a resurgence in popularity because virtually every piece of media coverage spun their story around the tale of the thirsty crow.

The term 'intelligent' comes to mind, but it's a term that has been mired in philosophical and scientific debate for centuries. Human intelligence alone can be difficult to define and we have the benefit of being able to explain why we did something a certain way; we can, however, all agree that humans possess the capacity for intelligent thought. Whether other animals have minds that endow them with emotion, thought, reason or consciousness is a much more controversial question, leading to disputes that have shaped the entire study of animal behaviour.

★ ★ ★

Darwin wasn't the first to write about animal minds. Nonetheless, his famously controversial proposal that differences in the minds of humans and other animals were of 'degree, not kind' sent shockwaves through his scientific and general readership, many of whom were horrified at the suggestion of any mental similarity. His writings on animal emotions inspired others to think about the mental capacity of animals, most prominently George Romanes, a respected physiologist who became fascinated by the possibility of intelligence in animals. Following in Darwin's footsteps, Romanes collected and catalogued hundreds of examples of animal smarts that had been shared with him, publishing the resultant *Animal Intelligence* in 1882. For Romanes and many others, there was no question that other animals possessed the capacity for intelligent thought.

The backlash came soon after, provoked by what critics saw as unscientific, excessive anthropomorphism. British psychologist Conwy Lloyd Morgan was prominent among them, emphasising the necessity of parsimony when considering apparently intelligent behaviour. He used his dog's behaviour as an example, pointing out that if someone walked past his house and saw his dog opening the garden gate, that person might conclude that his dog was an insightful problem-solver. What they didn't see, Morgan cautioned, were all the occasions on which he had trained the dog, progressively shaping its behaviour towards the goal of opening the gate. Insight was not necessary. Morgan set out a rule, stating that: 'In no case is an animal activity to be interpreted in terms of higher psychological processes if it can be fairly interpreted in terms of processes which stand lower in the scale of psychological evolution and development.' Known today as Lloyd Morgan's Canon, it remains one of the central principles in animal psychology.

Also in the 1890s, Russian physiologist Ivan Pavlov was studying how simple, automatic behavioural responses called 'reflexes' could be modified. In his most famous experiment, he trained dogs to associate the sound of a bell with the appearance of food. After the training, the mere sound of the bell was enough to cause the dogs to salivate. This type of learning, in which an individual comes to associate two unrelated stimuli in its environment, became known as classical conditioning. It was a breakthrough in the study of behaviour and found favour among early American psychologists, particularly John Watson who, in 1913, launched the field of behaviourism.

Behaviourists focused their attention on what was observable; i.e. behavioural responses that could be measured and quantified. Emotions and the workings of the mind were neither, so had no place in this field of study. The pendulum had well and truly swung away from the anecdotal approach of the nineteenth century; this rigorous, mechanistic approach had objectivity at its heart and it dominated American psychology for the first half of the twentieth century. Behaviourists believed that all behaviours could be shaped by an organism learning the effect of its actions, in the form of reward or punishment. They also believed that animals functioned independently of their environment, so to maximise experimental control they were tested in sterile cages, empty but for the stimulus and reward mechanisms.

Behaviourism yielded important insights: the laws that were established continue to underpin the study of animal learning. Nonetheless, for early twentieth-century naturalists, it was a lamentable way to study animals because it was so far removed from their natural behaviour. A different approach to animal behaviour emerged, primarily among European naturalists, called ethology. Like the behaviourists, early ethologists condemned the use of anecdotes to reveal anything about animal intelligence. However, their approach was at complete odds to the psychologists' artificial, laboratory-based studies of learning. Ethologists, including Konrad Lorenz and Niko Tinbergen, pioneered the scientific study of animals in their natural environments, with the use of 'ethograms' (catalogues of the natural behaviour of a species) to formulate relevant research questions. That wasn't the only difference. Ethology, having a biological basis, had Darwinian evolution at its core. This meant that the genetic basis of behaviour was of central importance – in complete contrast to the behaviorists, early ethologists studied hardwired behaviours such as instincts and imprinting to learn more about their evolutionary significance.

Behaviourism and ethology were opposites on a spectrum of behavioural biology and each side was extremely critical of the other's approach. It was not until the 1950s that they glimpsed common ground, with the realisation that both could benefit from the other. Today, behaviourism as a field of scientific study has largely been superseded by comparative psychology, although several key principles remain and are more widely used by teachers, animal trainers and others. Comparative psychologists retain a strong focus on controlled experimentation, solid experimental analysis and parsimony, but gone is the automatic dismissal of the cognitive function of animals. In turn, ethologists took on board the greater rigour and critical approach to behaviour analysis, and in the 1970s a synthesis between the biological and psychological approaches, called cognitive ethology, was pioneered to further the study of animal minds. Nonetheless, the behaviourists' opposition to studying animal minds resulted in decades of stigmatisation, with this research only gaining ground in the latter part of the twentieth century. Today, comparative psychologists and cognitive ethologists tackle similar questions, with their different perspectives continuing to provoke healthy debate about the interpretation of behaviour, as will become apparent throughout this book.

Intelligence is still a tricky term to use because without an accepted definition it's difficult to know what to measure; for that reason, researchers tend to talk about cognition or cognitive abilities instead. Cognition refers to the mental processes of acquiring, processing and retaining information, i.e. how an animal uses its evolved neurobiological tools. All animals have evolved ways to solve problems in their environment – including hardwired instincts, the capacity for environmental conditioning and cognition. They generally possess the cognitive abilities required for survival, which is why rolling out the same task to all species doesn't work[*] and why it's so important to understand a species' ecology and evolution to interpret its behaviour. Intelligence, it is suggested, is demonstrated if an animal applies its behavioural toolkit to a context outside of the one in which it evolved.

Aesop's crow dropped stones to raise the water level of the pitcher and the most cognitively advanced way to do this would be to draw upon knowledge about water displacement (i.e. that if an object is dropped into water it displaces the water by an equivalent amount), making a prediction about how to solve the problem without any trial and error. This is termed 'insightful' problem-solving (along the lines of Archimedes' or Einstein's famous 'Eureka!' moments) and in this context involves cognitive processing about the physical – as opposed to social – world.

Some of the earliest physical cognition research was carried out by German researcher Wolfgang Köhler, who in the 1920s studied chimpanzees at a research facility on the island of Tenerife. In one of his most famous experiments, Köhler suspended a much-desired banana out of reach and placed some boxes around on the floor. The animals initially tried to knock the banana down with a stick, but then seemed to 'hit

---

[*]There's a great cartoon that makes this point. A group of animals, including a monkey, penguin, elephant, bird and fish in a bowl, are lined up in front of a tree, and in front of them sits a man at a desk. He explains that in order to have a fair comparison, they must all take the same exam, which is to climb the tree. Of course, most of those animals are going to struggle, particularly the fish in its bowl. The cartoon was aimed at highlighting flaws in the education system, but it goes just as well for studying animal behaviour – and particularly intelligence.

upon' the correct solution to the problem. They stacked the boxes underneath the banana, climbed up and retrieved it. Köhler was impressed, writing that the chimpanzees had demonstrated insightful problem-solving. However, Morgan's Canon reminds us that before concluding complex cognition, simpler explanations must be ruled out – and for Köhler's chimpanzees this is not possible. The consensus today is that the chimpanzees' previous experience on different tests had already conditioned them to behave in certain ways, so that on the 'insight' trial they simply chained together these learned responses into one functional sequence. Although it is impressive problem-solving, it doesn't meet the criteria for insight.

More recently, an experiment that is analogous to Aesop's fable, known as the 'floating object' test, has been used to evaluate insightful problem-solving in other animals: initially orangutans and then the other great apes, capuchin monkeys and most recently elephants. The orangutans were presented with a tube that was attached vertically to the wire mesh of their enclosure wall; in it, a peanut floated in a small amount of water. To get the peanut, they needed to add water to the tube and, remarkably, the five animals tested all did so on their first trial. After trying unsuccessfully to reach the peanut, they went to the water drinker in their enclosure, took a mouthful of water and then spat it into the tube. They went between the water drinker and the tube as many times as was necessary to bring the peanut within reach. No training had been conducted with the animals and because they succeeded on their first trial the authors suggested that this demonstrated insightful problem-solving. A follow-up study with chimpanzees and gorillas showed some clear behavioural differences – the gorillas did not solve the problem, whereas some of the chimpanzees did. Most remarkable was the observation of one chimpanzee, who became frustrated that the water he was spitting wasn't going into the tube so started

urinating on it. Some of his urine went into the tube and at
that point he seemed to realise the effect it was having on the
peanut's position, so began directing his urine into the tube.
And yes, he ate the peanut.

The point is that to make conclusions about how an animal
solves problems, we need some certainty about what it has
already experienced and learned about the world. Aesop's
crow might have already tried to solve the task 20 times and
only after a stone accidentally fell into the jug did he learn the
effect it had in bringing the water closer (i.e. trial and error).
Or, he could have watched another individual dropping
stones, meaning that his attention was already drawn to these
objects (i.e. social learning – see Chapter 7). Based on a single
observation, it is impossible to conclude insight because it is
impossible to rule out these alternative, cognitively simpler
explanations. That's why a scientific approach to animal
problem-solving is necessary and that's what Bird and Emery
set out to do in their study. But before returning to that, it's
worth thinking a bit more about the rationale for studying
the cognitive abilities of birds.

* * *

For a long time, intelligence research was concentrated on
great apes. Much of this comes down to the fact that, before
Darwin's proposed branching tree of evolutionary history,
evolution was assumed to progress in a linear, hierarchical
manner called the *scala naturae*.* According to this idea, which
goes back as far as Aristotle, organisms could be ranked from
the least to the most evolved, with humans occupying the top

---

*Which subsequently became incorporated into the 'Great Chain of
Being', the belief that God had set out the social order for everything
on the planet, from rocks and minerals at the bottom (pure matter), to
angels and then God himself at the very top (pure spirit).

spot as the pinnacle of evolution. The rungs below us were occupied by chimpanzees and other great apes, then came the other primates, other mammals and so on, until you descended to the most primitive and 'least evolved' organisms at the bottom. Birds were thought to sit below mammals and were therefore considered to be evolutionarily and mentally inferior. The earliest studies of comparative brain anatomy only cemented this viewpoint.

Neuroanatomist Ludwig Edinger was inspired by Darwinian theory to work out how differences in brain organisation reflected evolution. Being grounded in the same progressive view of evolution, Edinger, together with many others in the late nineteenth century, expected that each vertebrate group would retain parts of the primitive brain and that brain complexity would increase in a step-wise, linear fashion that reflected the evolutionary ladder. To study this, Edinger and his student Cornelius Ariëns Kappers made thin slices of brains from different species, and compared both the relative size and shape of the main brain regions, as well as the appearance and distribution of neurons within them. Neurons, as a quick aside, are also called nerve cells and are the basic working units of the brain. Highly specialised, elongated* cells, their job is to transmit information, in the form of electrical impulses. Spaces between the ends of neurons, called synapses, allow the impulses to travel to other cells and connections to form. The more times that a particular neuronal connection is stimulated, the stronger it becomes and the faster information can be transmitted. The human brain has in the region of 86 billion neurons and it is estimated that each one can form up to 10,000 connections; that adds up to a gigantic amount of processing power and a huge capacity for learning.

---

*The longest neuron in our body is the sciatic nerve, a single nerve cell that runs from the lower back all the way down to our feet.

All vertebrate brains are made up of three major parts: the forebrain, midbrain and hindbrain. During embryonic development these areas diverge further, so the brain of a fully developed mammal looks different to the brain of a fully developed frog, although both are fundamentally divided into these three parts. Edinger's approach therefore made logical sense, and by sectioning and using cutting-edge staining techniques he found one striking difference between the brains of mammals and birds. It concerned the region of the forebrain called the telencephalon (or cerebrum), which is the most obvious part of the human brain, comprising two large hemispheres of wrinkly pale matter. In mammals, the telencephalon was formed of two distinct regions. First, a thin (2−4mm) outer covering of grey matter comprising just six layers of cells; second, an internal region in which the cell bodies are clumped together. Edinger and Ariëns Kappers could see that all mammalian brains shared the same basic organisation and that the outer layer was the most heavily folded in humans, giving the human brain its characteristic wrinkled appearance. Bird telencephalons, on the other hand, didn't have this layered outer wrapping. The implication was clear: the folded outer layer, which they named the neocortex, must have evolved after the bird and mammal lineages split, and because it was uniquely well developed in humans it was probably related to intelligence. Birds, Edinger concluded, had a reasonably well evolved telencephalon compared with other vertebrates, but without the neocortex they could only produce more primitive, instinctive behaviour. In their resulting publications on the evolutionary history of animal brains, they labelled all the parts accordingly, using the same terms to describe regions across animal groups that they believed were evolutionarily the same (or 'homologous'). It stuck for most of the twentieth century. Advances in the latter half, brought about by innovative new techniques in neuroanatomy as well as mounting evidence for bird cognition, cast doubt on the assumptions, but it was not until 2002 when the terminology for brain regions was overhauled.

Edinger was right that birds don't have a folded outer cortex, but since the 1960s there's been speculation that they may have a functional analogue. Proponents of the controversial idea focused on an area of the bird brain called the dorsal ventricular ridge (DVR). Cells in the DVR are arranged in large clusters called nuclei, drastically different to the layered cortex, which is why for a long time the DVR was considered to correspond to another, more ancestral region of the mammal brain. But molecular tools have revolutionised the field, enabling researchers today to make insights that Edinger and others could never envisage. A study published in 2012 found that marker genes expressed by cells in specific layers of the mammalian cortex matched up perfectly with those of the DVR, confirming that the two areas have similar developmental origins. The team also looked at the brain of a turtle and found yet another different structure containing the same marker genes. Evolution works in many ways, with functionally equivalent solutions evolving in animals that last shared an ancestor hundreds of millions of years ago. The latest evidence shows that no single neural structure is responsible for intelligent behaviour – and in any case the old ideas of mental inferiority are untenable in the face of the facts for bird cognition. The truth is, using 'bird-brained' as an insult is so twentieth century: if you want your insult to be aligned with current science, it needs to be a compliment.

Bird and mammal brains reflect different evolutionary pressures. Bird brains have been honed over time to enable efficient flight, with non-essential weight jettisoned; this means they are smaller and lighter than those of similarly sized mammals. A popular analogy, first suggested in 1999 by US psychologist Irene Pepperberg, is that the difference between bird and mammal brains is analogous to the difference between Macs and PC computers. The inputs to both are the same (sensory inputs), they both comprise similar-looking hardware (brains with three parts) and the outputs (behaviour) can look the same too. What's different are the processors and the internal wiring.

Birds are able to achieve complex cognition with smaller, lighter brains than mammals because they have more neurons per unit of brain tissue. For songbirds and parrots, it's twice as many neurons as mammal brains of the same mass; what's more, the 'extra' neurons tend to be found in areas of the forebrain associated with advanced cognitive ability. Parrots and corvids were found to possess the same, if not more, neurons in their forebrains than monkeys with considerably larger absolute brains. Within the birds there's variation too: a jungle crow's brain is about twice the size of a similarly sized pigeon. Corvids also have proportionally larger nidopalliums (the part of the DVR associated with cognition) than many other birds; for example, a rook's nidopallium is three times larger than a pigeon's and three times more densely packed with neurons.

The key question is why larger, heavier brains would evolve at all in some groups, because they don't come cheap. The brain is a hugely expensive organ, consuming around 20 per cent of the body's total energy. That means to evolve bigger brains, animals must either spend less energy on other activities (i.e. growth or reproduction), or they must increase their total energy intake. Intelligence, in other words, is costly. Bigger brains need to confer real benefits to make up for their substantial energetic demands – and that means helping the animal to overcome environmental challenges with which they'd otherwise struggle. In environments that don't change quickly or often, hard-wired behaviours may suffice; encoded in the genes, they enable the organism to do what it needs to with no expensive, superfluous brainpower. In environments that are complex or unpredictable, the cognitive ability to flexibly and quickly respond to new challenges may increase survival, meaning bigger brains can be favoured by natural selection.

In primates, a positive relationship has been found between brain size and social group complexity, lending support to what is termed the 'social intelligence hypothesis'. According to this, intelligence evolved to help individuals navigate the challenges associated with social life; we'll see more of this in the next chapter. The same doesn't really hold for birds, which typically form stable, monogamous pairs during the breeding season, and may either disperse or form larger flocks outside of that. This has led some scientists to suggest that larger brains may instead be associated with foraging-related abilities, such as cracking open hard-shelled, energy-rich fruits or using tools to extract food. There is evidence that across bird species areas of the DVR correlate with innovative feeding behaviour, but far more research is needed. At present, there is no single convincing theory to explain why some birds have larger brains than others, but behavioural flexibility finds favour with many, particularly when it comes to corvids.[*]

★ ★ ★

Wildlife biologist Bernd Heinrich started studying ravens in Maine in the late 1980s, conducting the first experimental study of insightful problem-solving in corvids with five hand-reared birds. Wild ravens occasionally infuriate ice fishermen by pulling up their lines from holes in the ice and stealing the bait or hooked fish. Heinrich's task was a version of this, with a piece of salami tied to the end of a string that was suspended

---

[*] The same is true of the parrots (*Psittacines*) and emerging evidence is showing that these birds are also capable of solving complex physical cognition problems.

from a perch. This simple task has been demonstrated in other birds – most notably pet goldfinches that had been trained to show the behaviour to the delight of their owners. But in all previous cases, the birds' behaviour could be easily explained by progressive learning – for example, the food was initially on the perch, and then suspended just below it and progressively lower. Heinrich wanted to know what his ravens would do without any of this 'shaping'. The results were mixed – some individuals seemed to 'get it', while others needed considerable experience. Heinrich struggled to publish his findings, submitting them several times before the paper was accepted. He was also shocked to discover that, although the idea of raven intelligence was widely quoted in the literature, there was only one other published study on their problem-solving (a study of counting, also conducted by Köhler).

At the time, animal cognition was completely, unashamedly, dominated by studies of primates and those interested in bird behaviour found themselves in a hostile environment. Clayton has spoken of feeling like a 'second-class citizen' while working on the neurobiology of food caching and spatial learning in the late 1980s. Apparently, in the hierarchy of scientific worth, those studying human brains were at the top, followed by primates, then non-primate mammals (especially rodents) and finally birds. Thomas Bugnyar, professor of cognitive ethology at the University of Vienna, has a similar story. He started studying ravens in 1996 and recalled needing to defend his PhD proposal to his advisory committee in Vienna. 'I was asked if I knew that I was not working on primates any more,' he chuckles. 'I was told that birds have a brain that does not allow for the kind of higher type of processing that I wanted to look for … This was the standing in the mid-1990s.' But a change was coming, an upswelling of interest stimulated by reports of corvids doing remarkable things – things that, according to the accepted dogma, they really shouldn't be capable of.

Heinrich's raven studies formed one strand of the new research, but there were several others. Among them were US researchers Russell Balda and Alan Kamil, who started working on Clark's nutcrackers in the 1980s. These birds store seeds in thousands of separate little sites during autumn and display remarkable spatial memories by recovering most of them several months later, even through a thick blanket of snow. John Marzluff started his PhD with Balda in the late 1980s; he says nutcracker spatial memory 'was the state of knowledge at that time'. Over at Cornell University, ornithologist Kevin McGowan banded his first American crow in 1989, initiating a long-term study that continues to this day. Then, in the mid-1990s, Bugnyar began his raven work, Clayton began her studies of memory and social cognition in the scrub jays and Gavin Hunt of Auckland University published a short report entitled 'Manufacture and use of hook-tools by New Caledonian crows', that shook things up even more. For Bugnyar, conducting his raven research alone in Vienna, it was a comfort to know that there was a growing international community who were fascinated by the same kinds of questions.

By 2004 the evidence base for corvid cognition, taken together with the reclassification of bird brains, led Emery, now Senior Lecturer at Queen Mary University of London, to coin the term 'feathered apes'. The proposal was that similar cognitive abilities had evolved convergently in apes and corvids, even though their common ancestor lived about 320 million years ago. Today, the field has broadened enough that corvids and apes aren't the only groups showing evidence of intelligence. But, Bugnyar emphasises, the term did its job in terms of getting people's attention and encouraging them to think outside the primate sphere: 'It was very important at the time that it was put forward. They're so strikingly similar. So that whole idea of convergent evolution I would underline and I still support this strongly.'

The adaptability and flexibility of corvids is clearly a big factor in their global success. Like foxes, it's why they thrive as well in towns and cities as they do in forests and moors. One of the things that comes up, time and time again, is how fast these birds can learn; for many researchers, it's a defining feature of the group. Corvids can learn after a single encounter and they can remember the details for years.

Marzluff, now professor of wildlife science at the University of Washington, has worked on many aspects of corvid ecology and behaviour, and has a particular interest in the human–corvid relationship. He had heard accounts of crows recognising and holding a grudge against people, including Konrad Lorenz, who wore a devil costume when banding jackdaws so that they wouldn't recognise him; and raven researchers in Alaska, who wore disguises to increase their chances of recapturing study animals. Intrigued, in 2006 Marzluff and students donned latex caveman masks and trapped seven crows around their leafy Seattle campus. When they later returned to the same spots, either wearing a different mask (Dick Cheney) or no mask, the crows paid no attention. When, however, someone returned wearing the same caveman mask, all hell broke loose. Marzluff has described how gangs of the birds would follow and divebomb them while making scolding calls. It seemed clear that they recognised the face and remembered the previous insult of being captured. Remarkably, it was not just the previously trapped birds that showed this response, other crows joined in too. Marzluff keeps testing them. In April 2020, 14 years after the initial capture, six of the 25 birds he encountered while wearing the caveman mask scolded him. Even more remarkably, 12 of 27 crows he encountered wearing the Dick Cheney mask showed an equivalent response. 'I think what is going on with Cheney is that there is some generalisation of masked people being dangerous,' Marzluff wrote to me. It's a fascinating development, given that Cheney shouldn't be associated with danger, and may set the scene for further tests. For now, though, Marzluff is waiting for the next phase of the study. 'I want to see reactions after all the birds that I tagged originally while wearing the caveman mask have gone. Currently there is one remaining (a female over 14 years of age), so I expect her to live only a few more years,' he said.

Corvids don't just recognise our faces. One study showed that hand-reared carrion crows paid more attention to recordings of unfamiliar voices compared with the voices of their owners. This is the opposite of what we'd see with dogs, which as domesticated animals respond more strongly to their owners' voices than those of strangers. For crows, the novel voice may represent a threat, hence their greater attention. They can even discriminate between languages. Large-billed crows in Japan were more vigilant when they heard the voices of strangers speaking Dutch compared with strangers speaking Japanese.

These findings tie in with another defining trait of the crow family, one that's a little more surprising – their fear of novelty (or 'neophobia'). Several studies have confirmed that corvids are more neophobic than other bird groups, and it's what makes them among the hardest birds to catch. Magpies, for example, are notoriously challenging subjects for studies in which they need to be repeatedly captured. For an opportunistic, generalist forager, it makes sense to exhibit some caution towards a novel object – it could, after all, be a disguised predator. And yet, this same fear response could inhibit them from discovering valuable new resources. This presents a bit of a paradox: how can corvids be both neophobic and innovative?

It comes down to learning. If a crow comes across a novel piece of litter in a city, for example, they will wait until another bird interacts with it first, quickly learning whether to categorise the object as safe or dangerous. Jackdaws, for example, are far more likely to interact with an object if they see another bird doing so. Ravens go further, being more likely to interact with an object if a bird with which it has a social relationship has been in contact with it first. In Marzluff's facial-recognition studies, the originally trapped crows learned that the caveman mask was dangerous, and their subsequent mobbing responses to that mask were attended to and informed otherwise naïve birds in the area. The latest finding suggests that the 'danger' category has widened to include Dick Cheney and possibly (although this is untested) any other novel mask.

Crow 'funerals' provide further information on how they learn about danger. Marzluff had previously observed birds gathering and calling around dead crows and was keen to investigate it more systematically. Swift, who had a 'wow' moment when she was researching potential research topics in bird behaviour and discovered corvid cognition, worked with Marzluff to define a PhD project centred on crow funeral behaviour. Her research, which involved masked volunteers holding taxidermied crows, as well as crow corpses in different positions being placed on the floor, indicated that the birds recognised the crows were dead and used the opportunity to learn about potential sites of danger and novel threats. As a slight – but I hope you'll agree forgivable – diversion, Swift says that one of her most exciting and unexpected findings was 'the tactile stuff', by which she means her discovery that some crows interacted with the dead birds – aggressively and sexually.[*]

* * *

---

[*] These atypical events tended to occur at the beginning of the breeding season and Swift concluded that 'alarm-induced arousal,' rather than sexual deviancy or any attempt to reproduce, is a likely explanation. The potential danger associated with the dead bird may have increased both aggressive and sexual arousal in the birds, and the mounting likely reflects the birds' inability to process these conflicting stimuli.

Back to Cambridge and Bird's rooks. They behaved as depicted in Aesop's fable, but did they 'invent' a new solution? As with all topics in cognition research, the question of 'can' an animal do something is straightforward to answer, the much thornier question is 'how'.

In fact, Cook and the other birds had taken part in a previous experiment in which they learned to retrieve food by dropping stones into a tube, although no water was involved. That means that coming into the Aesop's fable task the behaviour of dropping stones into tubes had already been reinforced. It's an important point and for some in the field negates the conclusions that can be drawn from their successful first trial performance. Maybe the rooks had simply learned that when faced with food in a tube and stones are available, they should drop them towards the food. This possibility doesn't invalidate Bird and Emery's findings, however it does provide more of a framework in which to interpret the behaviour.

In the conclusion to their report of the rooks' behaviour, Bird and Emery remind us that the moral of Aesop's fable was that 'necessity is the mother of invention'. Expanding on this, they proposed an alternative to account for the rooks' behaviour: 'It is cognitive generalisation that may provide the toolbox from which the solution could be drawn.' In other words, they proposed that the rooks could solve the task by drawing upon existing cognitive abilities such as rapid learning ability, concept formation and generalisation, which provided the tools to flexibly solve new challenges. For Bird and Emery, this generalised cognition was likely not unique to rooks, but present throughout the crow family. They threw down the gauntlet that other corvids may perform equally well on Aesop's fable.

The challenge was taken up by one of Clayton's then-doctoral students, Lucy Cheke, who was already working with Eurasian jays. These are the most colourful British corvid, but they're so secretive that although the RSPB

estimates there are 170,000 breeding pairs in the UK, they're
not commonly spotted. You may hear their raucous screech
in oak woodlands and catch a glimpse of chestnut as they
glide out of sight, particularly in autumn months when they
are busy caching acorns. Hand-reared birds, by contrast, are
another matter. When I visited the University's research
facility, a few miles out of Cambridge city centre, I was
introduced to each by name as we walked around, Cheke
chatting matter-of-factly to each bird and being chattered
back to in some cases. As she told me: '*Garrulus* [the Latin
name for the jays] means "excessively talkative". They are
chatty, curious and cunning. You walk into the room and
they look at you as if to say, "I see you. What are we doing
today? And where's my worm?"'

Cheke replicated the same conditions as Bird with her jays,
and like the rooks, they first had trials where they learned to
drop stones. The jays had no problem with the basic task, or
with the simple choice tests between a water-filled and sand-
filled tube, and between objects that would sink and float.
But in order to work out how they were solving these, Cheke
devised a trickier variation in which the correct behaviour
was confusingly counter-intuitive. In this 'U-bend' task,
three tubes stood vertically from a block of wood, each
containing water. The tubes at either end were wide, while
the central tube was much narrower – too narrow to drop
anything into – and contained the wax moth larva. The
central tube was in fact connected to one of the wide tubes,
while the other wide tube stood independently, but the
connection was hidden within the wooden base and so
invisible. That meant that to succeed, the birds needed to
drop stones into a tube that not only didn't contain the
worm but had no visible connection to it. Sounds impossible?
Cheke gave the birds one important clue – the correct tube
was always marked with the same colour, a seemingly

straightforward discrimination to learn. In fact, the jays struggled: they dropped objects at random into each of the wide tubes and showed no signs of improvement over the course of testing.

Children were also tested on the same experiment. Cheke found that those over the age of eight years learned which coloured tube to drop the objects into for their treat (a token that they could exchange for a sticker). As she explained: 'Children were able to ignore the fact that it shouldn't be happening to concentrate on the fact that it was happening. The birds, however, found it much harder to learn what was happening because they were put off by the fact that it shouldn't be happening.'

Cheke's study suggested that the jays were using a combination of trial-and-error learning together with basic understanding of the causal relationships at play. In the first two tasks, the birds could see the effect of their behaviour on the reward and they quickly learned which options resulted in food. In contrast, the correct behaviour in the U-bend task was causally counter-intuitive and this seems to have interfered with the birds' ability to learn what should have been an easy rule.

Rooks, you may recall, belong to the *Corvus* genus together with all crows and ravens. Jays, on the other hand, while belonging to the corvid family, are members of the *Garrulus* genus, which sits further up the evolutionary tree and is closer to the magpies of the *Pica* genus. That the jays also solved the task, therefore, implies that physical cognition evolved earlier in the corvid lineage than had previously been assumed.

One point has so far been neglected from the story – and it's an important one. Aesop's crow – as well as Bird's rooks and Cheke's jays – was undeniably using tools. The stone was an unattached environmental object, held by the bird and effectively oriented (i.e. dropped) into the container to alter the position of the desired reward. Rooks and jays are not natural tool-users, so perhaps this limited how far they could go with the task. How would a tool-using corvid fare?

Research on tool use in wild animals started with the work of Jane Goodall, who studied a group of chimpanzees living on a lakeshore in Gombe National Park, Tanzania (then Tanganyika). Shortly after beginning her study in 1960, Goodall observed a chimpanzee (whom she had named David Greybeard) poking a flexible stem into a termite mound. It was the first observation of tool use in a wild chimpanzee. Soon after, she returned to the same location and saw David Greybeard with another male named Goliath. She watched, mesmerised, as they stripped leaves off stems and probed into the termite mound. Writing to her boss – the renowned palaeontologist and archaeologist Louis Leakey – with her observations, he famously responded: 'Now we must redefine "tool", redefine "man" or accept chimpanzees as humans.'

We now know that all populations of chimpanzees make and use tools to achieve all kinds of things: digging into termite mounds, probing for termites, fishing for ants, dipping for honey, probing for bone marrow, cracking open nuts, pounding the 'heart' of palm trees to acquire syrup, sponging up water, cleaning body parts, hunting monkeys and throwing them as projectile missiles. The sheer frequency and diversity of tool use in chimpanzees sets them apart from other animals – however, they're by no means the only other animal to show this behaviour. What's more, it's not confined to primates, mammals or even vertebrates, as the following (non-exhaustive) summary should show. There are marine crabs that hold anemones in their claws for protection, wasps that use little pebbles to tamp down the earth outside their burrow, and veined octopuses that carry and shelter in coconut shells. Several bird species use tools, including green herons dropping 'bait' into water to lure fish; Egyptian vultures cracking open ostrich eggs by hurling rocks at them; and woodpecker finches, New Caledonian crows and Hawaiian crows probing for invertebrates with sticks and other plant material. Among mammals, elephants use branches for scratching and swatting flies; bottlenose dolphins hold sponges in their mouths to protect their rostrums (snouts) while foraging on the seabed; sea otters crack open shellfish by pounding them with stones; and capuchin monkeys crack open hard nuts by hitting them with rocks. All four species of great ape use tools, but they vary in the extent that this features in their natural behaviour. Chimps, as we saw already, use them routinely and in many contexts. Orangutans are also routine tool-users, but unlike the chimps this is mainly for extracting insects and opening fruit. Bonobos are more occasional tool-users, using objects as rain covers, scratching aids and sex toys,* while gorillas use them rarely.

Tool use captures our attention in ways that other behaviours do not, something that likely stems from its rarity (rare behaviours being considerably more interesting than things we see all the time) and its status for a long time as a defining human trait. We are the quintessential tool users; it is impossible to imagine a world in which humans evolved without using and making tools. For a long time, this was thought to be what separated us from the rest of the animal kingdom, and this was summed up in the influential theory of 'Man the toolmaker'. Goodall's report of tool use in chimpanzees shattered that theory, hence the gravity of Leakey's response.

Because of these associations, and the fact that chimpanzees are our closest relatives, it's long been assumed that tool-using

---

* Bonobo society revolves around sex – they form alliances and maintain these with sexual acts, even in same-sex groups; they engage in masturbation regularly; and they appear to be one of the few animal species that has sex for pleasure.

animals are more intelligent than their non-tool-using counterparts. Given our dependence on tools and the staggeringly complex ways in which we have used them to shape our world, it seems quite intuitive. But it's a risky belief, one that psychologist Daniel Povinelli and colleagues have termed the 'argument by analogy'. The basic assumption of the argument is that if we demonstrate a particular behaviour (and we know we're intelligent), then if another animal shows the same behaviour then it must also be intelligent. It's an easy trap to fall into because it's almost impossible not to view animal minds through the lens of our own. This is why early cognitive studies, which mainly involved apes and monkeys, were largely concerned with finding evidence of 'humanness' in the animals. Chimpanzees were brought up as infant humans, dressed in clothes and treated as family members to evaluate whether they would develop language or other human traits. Scientists' thinking has changed significantly since then; now, particularly for those with ethological backgrounds, good knowledge of a species' ecology and evolution is an essential first step to formulating research questions and developing appropriate experimental methods. The application of ecology to cognition hugely advanced the field.

Consider the chimpanzee and the gorilla. Both apes feed extensively on termites – insects that live in great 'cities' and construct some of the most remarkable natural structures known on our planet. Termite mounds, constructed using a mixture of soil, termite saliva and dung, can tower up to 10m above the ground. In order to feast on the termites inside, chimpanzees and gorillas use contrasting solutions. As Goodall had observed, chimpanzees insert tools and 'fish' for the insects. Gorillas, on the other hand, are powerful enough to simply break their way in. They don't need to use tools, but what does this reveal about their cognitive ability?

It's time to bust the entrenched myth that the mere presence of tool use in a species equates with intelligence. It's a rare behaviour, true, but automatically elevating the status of every animal ever seen to use a tool ignores all the important – and interesting – details. And as we know, the devil is in the detail. At one extreme, the little wasps that use pebbles to tamp down earth around their burrows are using tools in a very specific, stereotypical context – all evidence points to the behaviour being under tight genetic control. At the other extreme, some great ape and corvid species show habitual, flexible, creative manufacture and use of tools, and this is far more indicative of physical cognition. The point is that whether an animal uses tools is not the key question: what we need to explore is when and how they do so.

Rooks are not natural tool-users because they don't need to be. Out in the fields, grubs are plentiful if they just plunge their beaks deep into freshly ploughed earth, carrion is easy to come by and in any case, woodpeckers are found across the rook's geographic range so the ecological niche of extracting grubs from dead wood is already filled. And yet, bring rooks into captivity and they become tool-users. They have solved problems using stick tools, wire and stones, proving that they are more than capable if the situation requires it. Ravens do the same. Tool use is not important for their natural ecology, but bring them into captivity and they can learn after a single trial. For corvids, it does seem as if necessity is the mother of invention, supported by their possession of a cognitive 'toolkit' from which they can draw the correct solution. Nonetheless, it would still be revealing to evaluate the performance of a natural corvid tool-user.

Enter the New Caledonian crow, a bird that, at first glance, looks much the same as most other crows: it has the same glossy black plumage, chocolate-brown eyes and characteristic cock of the head as it surveys the world. But take a closer look, because the beak reveals something important: it's shorter, blunter and less curved than the beaks of other crows, and that's because it's evolved to hold and manipulate tools.

Endemic to the small archipelago of New Caledonia in the south Pacific, New Caledonian crows are the most prolific bird species to make and use tools. Studies only began in the mid-1990s, after Hunt reported how the birds crafted natural plant material into hooks to extract beetle larvae from dead wood. Before that, the scientific community had been unaware just how much tools featured in these crows' foraging behaviour. At that time, intentional tool-making was thought to occur only in chimpanzees and this lent weight to the ape-centred view of animal intelligence. Hunt's report had a huge impact. Wild New Caledonian crows are secretive and live in densely forested regions of the islands; nonetheless, Hunt and his colleagues (and subsequently others) have since filled in many of the blanks about these birds' natural ecology and behaviour.

At Oxford University, Alex Kacelnik, professor of behavioural ecology, and his then-postdoctoral researcher Jackie Chappell were intrigued by Hunt's findings. They initiated research with captive crows to investigate the birds' cognitive abilities. In one experiment, they varied how far away food was in an apparatus and found that the crows flexibly chose to use tools of appropriate lengths. In another, they found evidence for remarkable creative problem-solving. Kacelnik's graduate student, Alex Weir, ran this study in which a crow called Betty was faced with an out-of-reach bucket of food at the bottom of a vertical tube. Betty was provided with a straight and a hooked piece of wire to 'fish' for the bucket – the question was whether she would choose the functional hooked tool over the non-functional straight one? On the first test trial, Betty's mate (Abel, who had shown no signs of interest in the experiment to this point) decided to get involved and made off with the hooked tool. On the verge of stopping the trial after this apparent sabotage of

Betty's chances, Weir was astonished when Betty picked up the straight wire, jammed one end into the wadge of tape stuck around the bottom of the tube and bent it into a hook. She poked the hooked end into the tube and under the bucket's handle, pulling it out and getting her treat. Weir set the task up again, and again – each time Betty succeeded in retrieving the bucket by making herself another hooked tool.

The team was stunned, Kacelnik told me: 'We immediately realised that the anecdote was dynamite. At that time, no other non-human organism had been seen to make a hook as a response to a physical need.'

The experiment quickly became a textbook example of intelligent problem-solving. Further tests with Betty found that her modifications of tools were flexible – she bent aluminium strips using an adjusted technique that suited its different bending properties and would unbend these strips when she needed a longer tool. Although wild crows have since been observed bending the ends of stick tools when they craft them, Kacelnik does not believe that this negates the implications of Betty's behaviour: 'This is reassuring and expected: if they can spontaneously solve a problem in the lab, why wouldn't they also use that capability in the wild?'

New Caledonian crows continue to break new ground, astonishing the scientific community with their problem-solving abilities. From studies of captive individuals, we know that the crows don't just use stick tools to fish for food (i.e. they are not 'context-specific' tool-users, like the wasp). In one of my studies (during my DPhil with Kacelnik) we observed some individuals using tools for exploratory purposes, carefully prodding at a rubber snake and other novel objects before using their beaks to contact them, as we might do if we came across an unknown object in the woods. Another study from a different team reported crows poking tools into non-food objects and carrying them off, with one of the birds subsequently using the tool to prod the transported object under a groundsheet (where they often stash objects of value). Recent work led by Auguste von Bayern, who heads the comparative cognition group at the Max Planck Institute for Ornithology, together with Kacelnik and colleagues, showed that when some crows didn't have a long enough tool available they would combine two short tools to construct their own. One individual combined four components when an even longer tool was needed. The study was inspired by Köhler's finding that one of his chimpanzees, Sultan, spontaneously fitted two tools together to make a longer one – something that was taken as another example of insightful problem-solving. While it's a striking replication, the team are careful about using the term, pointing out that 'insight' is an unsatisfactory explanation because it doesn't shed any light on which cognitive processes may generate behaviour. Nonetheless, combining otherwise non-functional pieces to make a functional object demonstrates remarkable flexibility and suggests that the birds may be able to hold in mind the properties of the tool that they need – and that would be cognitively impressive.

Crows can combine tools together in other ways too. Another of my experiments showed that they could use little sticks to retrieve longer, otherwise out-of-reach sticks when the latter was necessary to reach food, with one bird successfully using a sequence of three tools to get the treat. Alex Taylor, associate professor at Auckland University, has expanded this line of research, investigating whether the birds will combine different tool types in a sequence. In one study, the birds completed an incredible eight-step sequence, involving three different kinds of apparatus. Like Köhler's box-stacking study, the birds were already proficient at each step. Nonetheless, chaining together this many steps in the correct order is impressive and, taken together with other findings, suggests that the birds can plan out their behaviour.

Taylor and colleagues have investigated several aspects of the Aesop's fable paradigm with New Caledonian crows. Unlike the rook and jay studies, none of the crows had prior experience at dropping stones to retrieve food, so it was unsurprising that no bird dropped stones into the tube on their first five trials. However, once the birds had learned the basic workings of the apparatus they completed variations of it with ease. Like the rooks and jays, they preferred to drop stones into water, not sand; they preferred to drop solid, rather than hollow, objects; and they preferred to drop objects into tubes with a higher starting water level. The experiment demonstrated that New Caledonian crows could quickly learn to use stones as tools, indicating a similar flexibility in problem-solving as for the other corvids. But on the causally confusing U-bend task, New Caledonian crows behaved the same as the jays: they dropped objects into each of the wide tubes at random, failing to learn the simple rule that would reliably lead to success.

The overall conclusion from these studies is that rooks, jays and New Caledonian crows can quickly learn to solve the Aesop's fable task, but only when they have the visual feedback to learn the outcome of their actions. The fact that jays and New Caledonian crows couldn't learn the colour rule in the U-bend task is perhaps surprising, given that rapid learning is a hallmark of the corvid family; however, it is consistent with evidence from other animals, including humans. Learning to associate a causally irrelevant cue with an outcome is harder than learning a relevant one, particularly when it conflicts with existing knowledge. For example, chimpanzees will quickly learn that the heaviest bottle in an array of otherwise identical bottles is the one that contains juice. But when all the bottles weigh the same and the one that contains juice is instead marked with an arbitrary colour, the animals struggle to learn the rule. It has been suggested that setting aside existing knowledge about object properties in favour of arbitrary, otherwise irrelevant, cues may be too cognitively challenging for non-human animals.

The Aesop's fable paradigm has most recently been tested with American crows. In Marzluff's lab at the University of Washington, PhD student Loma Pendergraft is investigating the responses of captive crows, but that's only part of his work. He's also using neuroimaging, aiming to bridge the crows' observable behaviour with their brain activity. Marzluff and colleagues previously pioneered the use of a neuroimaging technique called a PET scan with crows,[*] using it to investigate the neurological responses accompanying fear behaviour in the facial recognition study. They found that when crows that had been previously captured by someone wearing the caveman mask were exposed to a person in the mask again, there was activity in the regions of their brain associated with perception, attention, fear and escape behaviour, similar to our own fear responses.

Pendergraft has scanned the brains of every crow twice: on their first encounter with the Aesop's fable task and after they have been trained on it. The training process involves the birds first stretching into the tube to retrieve the reward (a highly desired Cheeto), then progressing to knocking stones

---

[*] Positron emission tomography (PET) scans are primarily used to detect cancerous tumours in people. An individual is first injected with a mildly radioactive form of glucose called FDG, which is taken up as an energy source and consumed by the body's cells. After some time, the individual is anaesthetised and placed into a scanner. The radioactive part of the molecule degrades into light-emitting photons, and the higher the readings in an area, the more glucose has been consumed. This either reveals the likely presence of a tumour or, in these experiments, greater neural activity.

off the tube edge into the water; after that, they're tested with stones on the table. He's still analysing the brain data – it's a complex and technical process – so it's a question of 'watch this space'. Whatever the results, it's an exciting step towards understanding more about what's actually going on when crows learn to solve problems.

The behavioural data, on the other hand, are clear. American crows seem to be less proficient at learning the task than other corvids, since only four of 16 birds progressed past the stage of accidentally knocking stones in. 'Most of the birds never got past this,' says Pendergraft. 'They just didn't really make the connection that "stone going in equals good".' Age seems to play a role in their performance, but in an unusual way. All the sub-adult birds got to the stage of accidentally knocking the stones in, but none progressed past this. Only the adult birds seemed to 'get it', which may reflect their more extensive learned experiences. But there were also some birds that categorically failed to get the food even when no stones needed to be dropped into the tube – and these birds were also adults. It doesn't seem to make sense but Pendergraft has a theory: 'Seems like adult birds were smart enough to completely figure out the task, or smart enough to realise that they don't need to solve the task, since they get daily food regardless.'

American crows have become proficient at exploiting the human landscape for food, rapidly learning about safe and dangerous people, rubbish bins, roads and other features of the human world. These are highly social, adaptable birds that live in communities and mate for life. Their performance doesn't seem to fit with the proposal of a family-wide generalised cognitive toolkit; however, as Marzluff suggests, the Aesop's fable task may not be the best one for tapping into the American crow's particular brand of cognition:

The ecological setting of a species is crucial to what it learns. Nutcrackers remember where they store their seeds. Pinyon jays remember social interactions with other jays. American crows remember people. New Caledonian crows craft tools. Ravens are experts at following cues of many species in their environment (including our gaze). None is smarter than the other. No task is a better indication of cognitive ability than any other.

It's an important point, although it's also true that exceptional insights about cognitive ability can come from animals solving challenges that are not typical of their ecology. To really know whether members of a species are capable of creative problem-solving, it can be useful to present them with tasks on which they can't generalise their typical species-specific abilities. 'We should not deny ourselves any path to wisdom,' Kacelnik notes. 'All animal competences combine prior specialisations with creative competences. What we need to understand is how they articulate.'

Fact or fiction?

Throughout history, crows have been represented in mythology, folklore and religion for their crafty nature, yet only in recent years have we started teasing apart the nature of their smarts. We now know that modern science supports the behaviour of Aesop's crow – and while we don't know if Aesop had ever witnessed such behaviour in person, it is probable that his existing knowledge about these birds influenced his choice of animal in this fable. Research has found that Aesop's crow could have been one of several possible corvid species,[*] which reveals an impressive ability to

---

[*] Admittedly, some of them would have been a long way from home.

learn about novel problems. We don't, therefore, need to seek an alternative character, although the Aesop's fable study has since been rolled out to other species, with kea, grackles and raccoons showing some success.

The mounting evidence on corvid behaviour suggests they may possess a 'cognitive toolkit' of traits that allows them to respond to environmental challenges flexibly and opportunistically. This is also true of chimpanzees, perhaps unsurprisingly given our relatively recent evolutionary divergence. But we'd have to go back around 320 million years to find our common ancestor with birds. The conclusion is that crows and apes have hit upon strikingly similar solutions to solve challenges in their physical and social worlds, using different brains that have been shaped by different evolutionary pathways. It's a fantastic example of evolutionary convergence and it's had the effect of breaking down the barriers that determined which animals should be studied for cognition. No longer were primates the only scientifically acceptable group. Today, the thriving field of comparative cognition includes a much more representative sample of species and, while there's still a long way to go, the past 30 years have witnessed profound changes. One exciting next step is to gain a better understanding of the neural bases of complex cognition. Neuroimaging techniques adapted for different animals, such as the PET scan used by Marzluff and team (and, as we'll see in Chapter 3, fMRI scans of dogs), have the potential to add several important pieces to the puzzle of how animals combine instincts, learning and cognition to solve problems.

The Aesop's fable paradigm is not without its critics. Povinelli, for example, has been outspoken in his disapproval of using a fable to structure scientific questions, while others have voiced similar queries about whether this test is the best method for investigating physical cognition in crows. On the other hand, the excitement surrounding the experiments has undoubtedly stimulated exciting new questions about animal minds. In turn, the fables have benefited from this research, with media interest causing a resurgence in popularity. No doubt Aesop would be pleased to know that his crow has flown from fiction to fact.

CHAPTER TWO

The Wolf in Sheep's Clothing

Once upon a time a Wolf resolved to disguise his appearance in order to secure food more easily. Encased in the skin of a sheep, he pastured with the flock deceiving the shepherd by his costume. In the evening he was shut up by the shepherd in the fold; the gate was closed, and the entrance made thoroughly secure. But the shepherd, returning to the fold during the night to obtain meat for the next day, mistakenly caught the Wolf instead of a sheep, and killed him instantly.

Takaya, a grey wolf, is thought to have been around two years of age when, in 2012, he dispersed from his pack on Vancouver Island. Traversing the city of Victoria (where he was spotted next to lakes, in parks and along a busy highway), the animal came to the easternmost point of the island and then made a remarkable decision: he took to the water and swam 1.5km to the tiny, uninhabited archipelago of the Chatham and Discovery islands. Local photographer Cheryl Alexander made it her mission to document Takaya's life on the islands and over the following years she developed a strong affiliation for the wolf. At less than 2km^2, his was the smallest wolf territory ever recorded, but he was the undisputed king of his palace: he learned to hunt seals, drink the contents of goose eggs and even dig 'wells' for fresh water. But he howled too, constantly through the mating season, either to advertise his territory or to advertise himself to any females in the area. Alexander has described how his 'lonesome howls' brought her to tears, knowing that they may have never reached the ears of another wolf. In January 2020, 10-year-old Takaya left the islands and swam back to shore where, fearing conflict with people, the authorities trapped and relocated him further inland. Unfortunately, he lacked both the skills to hunt large mammals and the necessary fear of people to survive in an area where hunting was rife – in March 2020, Takaya was shot.

Our relationship with wolves is complicated and Takaya's story illustrates multiple elements of this. The Songhees Nation, who own the islands, celebrated the return of a symbolic animal ('Takaya' is the word for 'wolf' in Songhees) and vehemently opposed attempts to capture him. Many, like Alexander, were captivated by the proximity and behaviour of this powerful icon of the wild. The draw of seeing Takaya, whose howls could be heard over the water, was too much for some people. One group, ignoring rules not to bring their own pets to the island, got a little too close for comfort when the wolf became curious about their dogs. The thrill of seeing him turned to fear and they called for emergency rescue, nearly sealing Takaya's fate there and then. But ultimately, it was a hunter that ended his life, just as hunters had done for tens of thousands of other North American wolves in the past hundred years.

Not many animals are quite as polarising as the wolf. And yet, without wolves, our lives would be altogether different. That's because these animals are the closest living relatives of our closest animal companions, domestic dogs, with whom they share 99.9 per cent of their genetic material. Sure, it might not look like it for many breeds, but that doesn't change their ancestry. Just how did it happen that dogs became man's best friend while their closest living relative became one of the most heavily persecuted animals on the planet? Like I said, it's complicated.

Around 45,000 years ago, when *Homo sapiens* arrived in Europe, their main competitors were Neanderthals and wolves. Neanderthals went extinct just a few thousand years later,* and while we continued to compete with this ancestral wolf species, something else began to happen too. The precise details about where or when are sketchy, but we're going back at least 15,000 years and potentially 20,000–40,000 years. The more interesting part is how and for many researchers this is best explained by a theory known as 'survival of the friendliest'. It suggests that some of the ancestral wolves started to scavenge around hunter-gatherer settlements, attracted by the smells of cooking, and by doing so they benefited from access to food scraps (and, according to a recent theory, energy-rich human faeces). These animals may have widened their territories to include the human camps,

---

*It is often assumed that *Homo sapiens* were directly responsible for *Homo neanderthalensis'* disappearance; however, it was an already vulnerable population, systematically fragmented by extreme climactic variations over the previous 100,000 years and suffering from inbreeding depression. The competitive presence of *H. sapiens* may simply have been a step too far.

unintentionally benefiting the people through their territorial behaviour. Not all wolves went down this path: most would have been fearful and aggressive towards people, so either avoided them or were killed if they came close. The ones that did well from scavenging were likely to have been less-dominant animals that benefited more from an alternative method of accessing food, as well as being friendlier towards people. The personality traits that predisposed these animals to be more comfortable around humans (i.e. lower fear of novelty and aggressiveness) would have been strongly advantageous to these wolves, leading to selection and the evolution of a new, friendlier kind of wolf. As we'll see in Chapter 5, pioneering experiments with silver foxes have revealed how selection for the single trait of tameness can lead to the rapid evolution of behavioural and physical traits associated with domestication. Wolves, it is thought, may have done something similar but driven by their own willingness to associate with people as opposed to human selection (as is the case for most domesticated animals). That is, they may have domesticated themselves. The canine that emerged became our loyal and constant companion: at our side to help hunt bigger prey, transport carcasses and protect against our enemies. The combination of human and dog made for a formidable and dominant predator.

Many faltering relationships can ultimately be explained by one or both parties changing, and this was certainly true of humans and wolves. The ancestral wolf species disappeared, survived by domestic dogs and a new lineage of modern wolves. This had likely already happened by the time, around 12,000 years ago, that hunter-gatherers were transitioning into pastoralists. Now, people formed settlements, grew crops and started to tend animals for meat, transport and labour. Sheep and goats were domesticated in the region of 10,500 years ago in the area of the Middle East known as the Fertile

Crescent*, followed closely by cattle. Wolves were unwelcome in these new communities, but unable to fully turn their backs, enticed by enclosures of livestock and the potential for an easy meal. Dogs became tasked with protection, pitted against their closest relatives and our long-standing adversaries. The conflict had well and truly begun.

★ ★ ★

'The Wolf in Sheep's Clothing' is not the only one of Aesop's fables to paint the wolf as a cheating tyrant. In 'The Wolf and the Lamb', the wolf promises it won't eat the lamb and then, like a true despot, goes back on its word. In 'The Wolves and the Sheep', the wolves persuade the sheep to dismiss the pack of dogs that protect them, with a false promise of peace and friendliness. As you can probably guess, the wolves feasted on the entire flock. In 'The Wolves and the Sheepdogs', it was the dogs that the wolves deceived, persuading them that instead of being slaves to men, they should be brothers and live together, before turning on them. And in 'The Wolf and the Shepherd', the wolf played the long game, hanging peacefully around the sheep for long enough that the shepherd eventually left them in his charge. On his return, of course, he didn't have many sheep left. There are others, but it's clear that there's a bit of a theme. Indeed, in the couple of fables where the wolf gets outwitted by its prey, it laments that it did not stick to its trade of butcher.

And yet, the ancient Greeks also held wolves in high esteem. Their ferocity and strength could be respected, even held up to represent ideal warrior qualities. Some of the gods

---

* Also known as the 'Cradle of Civilisation', this semi-circular area of the eastern Mediterranean was named for its rich soils.

were associated with wolves, such as Apollo, son of Zeus, who was known as the Lord of the Wolves. He was a protector and the wolf his divine messenger, sent to ward off enemies and save the young, old and weak.

The wolf also had an important place in ancient Roman culture, being a central figure in the legend of its founding. It's said that the baby twins, Romulus and Remus (fathered by Mars, the God of War, himself strongly associated with wolves), were cast out of their kingdom by a jealous, fearful king. A she-wolf found them and nurtured them until they were discovered by a shepherd. Romulus and Remus went on to establish the city of Rome and the she-wolf became a legendary symbol in Roman culture, depicted on Roman coins as early as the third century BC; the imagery was even used as the emblem of the 1960 Rome Olympics.

Yet, as the centuries passed and more people moved to the great cities of the ancient world, the balance started to tip against wolves. In Europe, the spread of Christianity sounded the death knell: wolves were painted as the embodiment of evil and, most damningly, the enemies of Christ, meaning that ridding the world of them became something of a religious obligation. We can get a sense of how poorly wolves were thought of from medieval bestiaries, which are a bit like early medieval natural history encyclopaedias, except that all the 'facts' about the creatures included relate to their place in the Christian worldview. Here's an excerpt from a twelfth-century volume:

> Wolves are known for their rapacity, and for this reason we call prostitutes wolves because they devastate the possessions of their lovers. Moreover, a wolf is a rapacious beast, and hankering for gore ... The devil bears the similitude of a wolf: he who is always looking over the human race with his evil eye, and darkly prowling round the sheepfolds of the faithful so that he may afflict and ruin their souls.

The downfall of wolves came when religious beliefs became enshrined in law. The tenth-century King of England Edgar the Peaceful is said to have demanded tax payments in the form of wolf heads and skins. Then, in 1281, King Edward I ordered the extermination of all wolves in England. It's generally accepted that by the sixteenth century, wolves were extinct. They'd already vanished from Wales, but were able to hang on in Scotland's great highland forests until the 1600s[*] and in Ireland until the mid-1700s. There have not been wolves in the British Isles since.

It was the same story in Europe: by the late eighteenth century, wolves had been driven out of France, Germany, Italy, Spain and the low countries. Irrational fears of the animals grew, spurred on by folk stories in which wolves were driven not to kill livestock (as depicted by Aesop), but people. 'Little Red Riding Hood' is a prime example,[†] in which the wolf eats Grandma and then disguises itself in her bed to deceive and eat the girl. It's basically Aesop's wolf, except here it wears Grandma's clothes instead of the sheep's fleece.[‡] The malevolent wolf of Prokofiev's 'Peter and the Wolf' is another case in point, terrifying people into believing that vicious creatures lurked in the darkness, waiting to catch them should they stray too far from safety. The seeds of fear

---

[*] According to official figures; there are unofficial reports of wolves surviving until the eighteenth century.

[†] It was Charles Perrault who wrote down this folktale in the seventeenth century; however, it is thought to have its roots in an eleventh-century poem.

[‡] The Italian version of this story was called 'Grandmother in Disguise' and was far more horrific. After eating the old woman, the wolf saves some of her flesh and blood, which he subsequently invites the girl to eat and drink, before he requests that she take off her clothes and join him in bed. In this case, the wolf embodied a different form of evil for young girls.

had been sown and fear, once implanted, is very difficult to remove.

Europeans took their fear of wolves to the New World too, unmoved by the positive ways in which the animals were represented in native culture and folklore. In 1607, when British colonialists arrived in Virginia, wolves had all but vanished from their final refuges in the Scottish Highlands. Yet here they were, widespread across the vast wilderness of North America, a danger to the colonists' imported livestock and another part of this hostile land to tame. The first official act of wildlife control came in 1630, with a bounty on wolves in Cape Cod, Massachusetts, and as the colonies expanded so did the anti-wolf sentiment. Shooting, trapping and poisoning were common, and the promise of generous bounties even led some men to take up new occupations as full-time 'wolfers'. The Montana Historical Society notes that wolves were plentiful in the newly created Montana Territory in the 1860s. Then the fur traders and bounty hunters moved in, working to the simple motto that 'the only good wolf is a dead wolf'. Between 1871 and 1875, an estimated 34,000 wolves were killed in the area of northern Montana and southern Alberta alone. By the late 1880s, total extermination of wolves became the goal, even for those involved in wilderness protection. The biologist William Hornaday, a pioneer of the American conservation movement and instrumental in the preservation of American bison, was scathing in his 1904 work *The American Natural History*, writing: 'Of all the wild creatures of North America, none are more despicable than wolves. There is no depth of meanness, treachery, or cruelty to which they do not cheerfully descend.' In 1915, the US Congress authorised the elimination of any remaining large predators and that same year the US Biological Survey hired its first federal wolf-hunters. When they were disbanded in 1942, they had killed more than 24,000 wolves.

There were efforts to portray wolves differently: Rudyard Kipling's *The Jungle Book* stands out for his remarkably true-to-life portrayal of wolf family groups; Jack London's *The Call of the Wild* and *White Fang* portrayed wolves favourably too. And Ernest Thompson Seton's *Lobo, the King of Currumpaw* provides an evocative account of Seton's relationship with a feared male wolf named Lobo. More recently, greater environmental interest and awareness means there's more of a market for alternative portrayals of animals – for example, we now have fairy tales in which the wolf is the victim, not perpetrator, of wickedness. But it is still the case that most stories, cartoons and films paint the wolf as the villain – and that's the problem. A long time ago the persona of the Big Bad Wolf seeped into our collective subconscious and every reinforcement of it strengthens that belief. The sad outcome is that for many people, the lines between fact and fiction became too entwined for them to really know the truth. That's where science comes in.

Wolves belong to the evolutionary family called the *Canidae*, along with foxes, jackals, dingos, coyotes, dholes, maned wolves, racoon dogs, bush dogs and African wild dogs. There are 35 species of canid living today. Going back in evolutionary time, the *Canidae* was the first lineage to split from an ancestral group of carnivores that subsequently diverged into several other families: bears, racoons, weasels and the aquatic predatory group that includes seals, walruses and sea lions. That split happened more than 40 million years ago, making the canids the most ancient living family.

The grey wolf is the largest species, with a maximum body length of about 1.6m and average weight of 50–70kg. The smallest is the fantastic little fennec fox, a desert-dwelling canid that weighs just over 1kg, measures about 40cm and is characterised by its disproportionately large ears (up to 15cm). Canids are hugely successful worldwide, represented in every

continent except Antarctica. Historically, the grey wolf was
the world's most widely distributed mammal; today, however,
their original range has been reduced by about a third,
particularly in parts of western Europe, Asia, Mexico and the
US, where they've come into conflict with humans for preying
on livestock. Wolves are mainly now confined to the remote
wilderness regions of Canada and Alaska, as well as northern
areas of the contiguous United States, Europe and Asia.

There are three key aspects to Aesop's fictional wolf, along
with the big bad wolf of Grimm's fairytales and others. First,
and most obviously, they are voracious killers; second, they
are loners; and third, they cheat and deceive their hapless
victims. Let's look at the facts.

Grey wolves, as large carnivores, must hunt and kill other
animals for meat. They mainly prey on hoofed mammals
('ungulates'), such as elk, white-tailed deer, caribou or bison,
although smaller mammals such as beaver and snowshoe hare
are frequently taken too. As pursuit hunters, they are
wonderfully adapted with long, slender legs and specially
adapted 'wrist' bones, which help them to run fast and chase
down prey over long distances (anywhere from 100m to
more than 5km). There's a common misconception that
wolves are outstanding hunters, which is understandable
given the fact that a pack is capable of taking down huge
prey and that, unlike us, they do it with their teeth. In fact,
on average only 14 per cent of attempted hunts are successful,
although this varies according to location. When they do
make a kill, wolves tend to take out the most vulnerable
individuals in a group: the very old, very young or injured
individuals that don't have the stamina. Fit and strong moose,
bison and deer are likely to get away, and since only a
minority of the prey population is made up of very vulnerable
individuals, wolves must work hard for their food. So, killers
they may be, but that's a fact that comes down to their
biology and their need to eat, rather than them having a
thirst for killing.

Being the largest canid, the grey wolf is the most dangerous to humans and attacks do happen. In Europe and North America, just a handful of fatalities have been reported over the past decades. It's a different story in India and Pakistan, where wolves have contracted rabies from feral dogs and where the combination of habitat loss and the lure of livestock has forced them into uncomfortable proximity with human settlements. Here, wolf-related fatalities number the hundreds. However, according to the Wildlife Institute of India, this is still exceptionally rare given how many opportunities wolves have to attack people; they concluded that attacks should be considered an aberration of natural wolf behaviour.

Next, the frequent portrayal of wolves as loners. Today, 'lone wolf' is a well-known turn of phrase, often applied by the media to individuals that society considers 'odd' for keeping themselves to themselves, and frequently in the aftermath of a serious crime. It's also a misnomer when it comes to wolves, because Kipling was spot on with the key to the wolf being the pack. While packs of up to 36 individuals have been reported, the normal size is between five and 12. Wolf pups are usually born every spring and stay with their parents for anywhere from 10–54 months, which means that a pack might contain several generations of offspring. Because pups reach adult size by winter, a pack may appear to comprise all adults, whereas in fact many of them will still be youngsters. We'll come back to pack composition shortly. Each pack defends a territory, which can range in size from 75–2,500km^2, depending on how much prey is available. Prey characteristics also influence the number of pack members: packs are bigger in areas with large ungulate prey, such as moose and bison, compared with packs that prey on deer or scavenge human waste. Studies of the wolves in Yellowstone National Park, for example, found that bison are three times more difficult to kill than elk, and scientists have found that packs that successfully prey on bison were considerably larger than packs hunting elk.

Defence is an important aspect of pack life and territories are patrolled frequently and diligently. Much of a wolf's time is spent walking, covering the same trails around the territory for many hours of the day. By repeatedly scent-marking along the boundaries, they're communicating to other wolves that the area is taken and actively defended. Howling is also used to communicate between, as well as within, packs. Each wolf has its own distinct voice, allowing scientists to tell individuals apart, and when they howl as a pack, which they do several times each day, it's a soulful, ominous chorus. Howls can travel for up to 16km over calm, open spaces and they serve an important function, helping to prevent packs 'bumping into' each other. When packs meet, vicious and often deadly fights occur. Data from Denali National Park in Alaska, where wolves have been monitored since 1986, show that conflict with other packs is the primary cause of death among wolves, generally in winter when the animals are forced to search for food outside their normal territorial boundaries. It's the same in Yellowstone, where wolves were reintroduced in 1995: the number one cause of death for wolves within the park is other wolves (while outside the park it's humans).

Why, then, do so many stories feature a solitary wolf? I put it to Dave Mech, American wildlife scientist and legendary wolf biologist, who says it's a misunderstanding about wolf biology: 'It is transitional. They are not real loners, but appear alone.' It comes down to two key facts. First, wolf packs are not fixed over the course of a year. In winter, when the environment turns harsh and food is scarce, being in a pack is hugely important for survival. In the summer, wolves do their own thing a bit more. At this time, the breeding pair have a den and pups, and the non-breeding wolves may travel around on their own or in small groups to hunt and patrol. They return to the den area each day to provision the pups with food, but you're far more likely to spot a lone wolf at this time. Second, when young adults disperse from their family group, like Takaya, they are likely to be alone. This happens between the age of one and three years, when they strike out to find a mate and establish a territory. They can travel hundreds of miles to find these things, which may take several months. Takaya's life, it should be acknowledged, was not typical for a wolf.

The third query is whether wolves are capable of deceiving others, and that is the crux of this chapter. To answer it, we need to have a good look at how wolves interact with each other. Extensive field research with large, free-ranging populations across (predominantly) the USA and Canada has provided important insights into wolves' natural behaviour and ecology, while studies of captive animals have focused on developmental and cognitive aspects to their behaviour. Both have their roots in pioneering research conducted over 85 years ago.

In 1934, Rudolph Schenkel initiated his studies of captive wolves in Switzerland's Basel Zoo. Over more than 10 years of study, Schenkel made detailed observations of two packs: each had up to 10 wolves and occupied an area of approximately $200m^2$. He published his findings in a book entitled *Expressions Studies on Wolves* (1947). At this time, no thorough study of wild wolf behaviour existed, so Schenkel referred readers to the fictional works of London and Seton for an 'approximate picture' of wolf natural history. His observations of captive wolves revealed that their behaviour followed a broadly cyclical pattern over the course of a year. Things started to heat up each winter, beginning with an increase in 'friendly relations' and followed by violent rivalries and pair formation. The pack is formed, Schenkel concluded, when the breeding pair establish themselves at this time. He introduced the term 'alpha' to describe the top male and female, proposing that 'by incessant control and repression of all types of competition (within the same sex), both of these "α-animals" defend their social position.' The implication was that individuals in the pack were constantly vying for dominance, but were held in check by the alpha pair, who exerted their authority by eating first, preventing other individuals from breeding and aggressively reminding their packmates who was boss. Alphas, in other words, were bullies. Schenkel drew frequent parallels between the wolf behaviour he observed and that of domestic dogs, meaning that these studies informed subsequent treatment of and research into both groups. The problem is that while Schenkel's interpretations of social structure characterised what he was seeing, they did not provide an accurate representation of natural wolf behaviour, coming from a mix of animals from different packs and different locations that had been lumped together at Basel Zoo. Schenkel was aware of this and cautioned that the animals were not able to exhibit some of their natural tendencies, such as dispersing during the breeding season. Nonetheless, his ideas took hold and even today, many stories and films portray a wolf pack that is led by an alpha male.

The confines of a 200m² enclosure do not invalidate Schenkel's research – he provided the first in-depth descriptions of individual wolf behaviour. But using it to draw conclusions about the natural behaviour of an animal whose territory size may be anywhere from 75–2,500km² was always going to be problematic. To really understand how wolf packs work, wild animals needed to be observed. And just a few years after Schenkel started, the first studies of wild wolves began too.

American biologist Adolph Murie was hired in 1939 to study the relationship between wolves and Dall sheep in Mount McKinley National Park, Alaska (renamed Denali National Park in 1980). A catastrophic decline of the sheep population was blamed on wolf predation and Murie's job was to gather the evidence. Instead, he found that the sheep population had crashed following severe late-winter weather, absolving the wolves. Not only that, Murie's data indicated that wolves played a crucial role in the ecosystem, something that was completely at odds with the ethos of wildlife management at the time. His report was published in 1944 and, fortunately, the park managers took it seriously, scrapping predator control in the park.

Mech was inspired at an early age by Murie's research and particularly by the realisation that one could make a living by tracking animals in the wilderness. As a student, he tracked and tagged black bears in the Adirondacks, a dangerous and exhausting task that was a world away from the easier, safer methods afforded by GPS and dart guns today. Still, Mech remembers those as the good old days of field research. In 1958, he began his PhD research on wolf–moose relationships in Isle Royale National Park, a 2,200km^2 area of islands in north-western Lake Superior. He gathered most of his data from aerial surveys aboard a light aircraft, supplementing it with information gathered on foot; however, it became clear from the latter just how elusive the animals were: only three wolf sightings were made from over 1,400km of hiking over three summers. In his 1966 write-up of the study, Mech reflected on the plight of wolves: 'The anti-wolf prejudice of most of us was instilled when we were naive and innocent tots.' He commented that among the earliest songs learned in a child's life was 'Who's Afraid of the Big, Bad Wolf?' and that this was followed by the stories of 'Peter and the Wolf' and 'The Three Little Pigs'. It was little wonder to Mech that wolves avoided people like the plague.

When Mech started out, Schenkel's ideas about alpha status were thoroughly entrenched and Mech accepted them too, popularising them in his book *The Wolf* (1970). Referred to as the 'bible for wolf researchers', the concept started to be applied to humans – many of us will find familiarity in the concept of the alpha male or female. Yet, as Mech continued to study wild wolves, he started to question his own claims. During the summers of 1986–2010, he conducted field research on Ellesmere Island in the high Arctic. Unlike Isle Royale, the wolves here were so isolated from people that they had no fear, allowing Mech to habituate them to his presence and observe them from as close as 1m away. His detailed observations were at odds with the seething hotbed of dominance disputes and squabbling portrayed by Schenkel; rather, his observations showed packs to be a family affair. The 'alpha' pair haven't fought their way to the top of the heap, they're dominant because most of the animals around them are their offspring. The breeding pair is the core of the pack and, since most mated pairs stay together for life, it's highly stable. Unlike Schenkel's unrelated wolf groups, Mech did not observe any instance of dominance being contested during his 13 summers observing natural packs. That's not to say there wasn't any aggression between pack members, just that individuals weren't constantly vying to be top dog. In fact, pups establish their relative dominance status in the first weeks of life and communicating this is an important component of their behaviour.

Body posture communicates a lot of information for all kinds of animals. This was clear to scientists long before studies of wolves began in the wild or captivity. Darwin, for example, described in characteristic detail the differences in body posture between a hostile and a friendly dog. Watch your pet dog the next time it hears a car pull up and someone walk up to your front door; then see how it changes when it realises it's not a stranger but a family member. There are clear postural differences and it's exactly the same for wolves. A dominant animal is a model of tense alertness, standing tall with erect ears and its tail stiffly held up from its body. A subordinate animal, on the other hand, does all it can to be the opposite: cringing towards the dominant individual it lowers its body, flattens its ears and curls its tail between its legs. Mech observed that within the pack all individuals, including the breeding female, adopted submissive postures towards the breeding male, while all offspring adopted submissive postures towards both breeding adults. The only exception was when the female had pups, when she assumed temporary dominance over her mate. Mech observed the male approaching the female's den with a submissive 'wiggle walk', animatedly waving his tail and back end in the manner of a subordinate, before regurgitating food for her and the pups.

In an influential publication in 1999 entitled 'Alpha status, dominance, and division of labor in wolf packs', Mech renounced Schenkel's concept of the alpha wolf, drawing the analogy of studying people in refugee camps to learn anything about human family dynamics. He wrote: 'The concept of the alpha wolf as a "top dog" is particularly misleading.' His updated theory was well received by scientists and today it would be unusual to hear a scientist describe the composition of a pack in Schenkel's terms. In other groups, he explains, it's still entrenched: 'Many naturalists and others dealing primarily with the public continue to use it, as does the media, which has not caught up with the science.'

Observations of wild wolves show that pack members communicate and cooperate, particularly when it comes to hunting. On Isle Royale, Mech observed a line of wolves heading along the frozen shoreline before suddenly stopping and facing upwind towards a large moose. After a few seconds, 'the wolves assembled closely, wagged their tails and touched noses'. They then started towards the moose. Of course, we can't know what passed between the wolves, but Mech's observation offers a tantalising glimpse into the form this might take. It also speaks to the bonds that exist between members of a pack.

The late biologist Gordon Haber studied wolves in Alaska for 43 years, and his observations demonstrated both how strong their bonds are and the devastating impact of dismantling pack structure. As you may recall from Mech's early research, the wolves of Denali frequently predate Dall sheep, which is not an easy task as the sheep are far better at climbing, so can evade the wolves by running uphill. One pack that Haber studied, the Toklat pack,* had become Dall sheep specialists, learning how to wait and ambush the sheep from above. Haber's observations indicated that young wolves required two to three years to learn how to become proficient hunters. In the winter of 2005, the Toklat pack's breeding female was caught in a snare. Two other females from the pack stayed with the trapped female, who was likely their mother, and were subsequently trapped as well. After the deaths of these females, the breeding male returned for months to the same spot until he too was killed by a hunter. The six remaining wolves were all juveniles and they lacked the knowledge and experience to hunt Dall sheep; forced to turn to alternative prey, such as snowshoe hares, they lost an important survival skill. This, together with other examples where the loss of a breeding animal had knock-on effects for the rest of the pack, led Haber to dismiss the management strategy of culling a certain number of individuals, calling it 'ecological nonsense'. It's not how many animals are killed, he argued, but which. Haber died when his plane crashed over Denali during a routine survey, but his arguments continue to be supported. A recent study of the Denali wolves, using data from 1986–2013, found that 77 per cent of pack dissolutions were preceded by the death of a breeding animal.

---

*The Toklat pack (aka 'East Fork pack') was thought to be the oldest known and most studied wolf pack in the world, with data stretching back to Murie's studies in the 1930s. Today, the pack is feared to have been lost, the result of renewed predator control in Denali.

Supportive data are also available from the Yellowstone wolves, where 41 Canadian grey wolves were reintroduced between 1995 and 1997, after being absent since the late 1920s. They settled in well and since 2009 the population has been reasonably stable at around 100 individuals. It is arguably the most intensively studied population in the world. Doug Smith leads the Wolf Restoration Project and a key finding for him has been learning that wolf packs are matrilineal – i.e. the territory is passed down from mother to daughter, or sister to sister. Young males leave the pack to avoid inbreeding, while daughters seem to stick around for the opportunity to step into the breeding role.

Another surprising finding from the Yellowstone wolves is that although absolute pack numbers are important in predicting the outcome of fights, they aren't the most influential factor. By analysing the outcome of wolf pack fights, researcher Kira Cassidy found that age had a stronger effect. A pack with one more 'old' wolf than the opposition had 150 per cent greater odds of winning a fight, which seems counter-intuitive because these wolves are past their physical prime and less likely to participate. Yet, as Cassidy has written, 'what old wolves possess is experience'. That old wolf has encountered conflict many times and has survived; it has experienced the death of family members and had the opportunity to learn when the pack should fight and when it should concede. Life experience is valued in wolf society, something that we could perhaps take note of.

Long-term studies of wolves have, and continue to, put many of the key pieces together, providing a solid grounding on population biology and natural behaviour. Detailed descriptions of wild wolf behaviour suggest that these are intelligent animals, yet to get an idea of what may be going on in the mind of the wolf, we turn to studies of captive animals.

The earliest wolf cognition studies asked whether wolves and dogs differ in their use of human communicative cues, and we'll come to that in the next chapter. Today, there are several research groups investigating wolf cognition around the world, with the Wolf Science Center (WSC) near Vienna, Austria, providing many important insights. Established in 2008, the WSC's mission is to compare the cooperative and cognitive abilities of dogs and wolves[*] reared under identical, pack conditions. Friederike Range, associate professor at the Konrad Lorenz Institute of Ethology in Vienna, and one of the founders, explains more. 'Wolves have been very much hunted, so only the shyest animals survived. A big difference between wolves and dogs is that wolves have this inherent fear of humans. For dogs, of course, we selected against this.' And yet, at the WSC the wolves are willing participants: neither fearful of the researchers nor dependent on them, these wolves view people as cooperative partners. It's all down to the rearing process, which is identical for wolves and dogs. Human caregivers start interacting with pups at the age of 10–12 days, before they have even opened their eyes. This ensures that the caregiver is among the first things that a pup sees, which is particularly important for the wolves. It takes round-the-clock attention for the first five months of life to ensure that the young wolf becomes fully comfortable around people and differences between the animals are seen early on. 'The wolves are far more active,' says Range. 'They're more explorative and much more interested in their environment. And they're more persistent from the beginning – persistence is a huge difference between the two.' Differences are also seen overnight, with the young wolves more inclined to wake their human caregiver by playing or going to the toilet on them. Dog puppies tend to sleep the night through, just like their humans.

---

[*] Canadian grey wolves are used rather than European grey wolves, because they are apparently more relaxed around humans.

After a few weeks, the animals start to learn commands such as 'sit', which helps with subsequent training. Then, at around five months of age, the pups are integrated into one of the wolf or dog packs at the park. Dogs, of course, don't need to be reared in such a way, but by doing so the team are controlling for the possibility that any differences seen in their later behaviour result from different experiences in early life.

The WSC have found some fascinating differences and some important similarities. Important, because they've helped to debunk the long-held belief that the behavioural traits of dogs are unique and special. Although it came to be accepted that wolves were dogs' closest living relatives (an early hypothesis had suggested coyotes), there has been resistance to the idea that they possess similar cognitive traits. Instead, it has been proposed that human-induced selection pressures on dogs for particular social skills, such as attentiveness or cooperative ability, led to their evolution as uniquely adapted animals. In fact, the more comparative studies that are done, the less credible this idea is.

There are also some clear differences. Free-ranging dogs behave very differently towards each other than wolves do; for example, when sharing food. In a wolf pack, all members get to eat and while there are spats between individuals around a carcass they tend to reconcile afterwards, like chimpanzees, ravens and other highly social animals. For the pack to hunt together they need to be a cohesive unit, which means it's important to maintain and repair social bonds. In a free-ranging dog pack, the most dominant individual monopolises food and lower-ranking dogs avoid it to reduce conflict. After fights, one study showed that dogs actually avoided each other, indicating that thousands of years of partnership with humans has eroded this key pack survival skill. What's more, wolves will choose to provide food to a member of their pack who does not have any (but not another outside of its pack), while dogs will not.

Another key difference is seen in problem-solving ability. This was assessed using what's known as the 'rope-pulling task', which is the standard test of animal cooperation. In this, an individual is paired up with another from their group and faced with a food-retrieval problem. A tasty treat is placed on a platform, which is out of reach of both animals. Around the platform edge, a rope has been threaded and the loose ends dangle into the testing enclosure. To drag the platform in and access the food, each animal must pull one end of the rope and crucially, they must pull at the same time – if only one of them pulls, the rope is simply unthreaded from the platform. Wolf pairs worked together perfectly: they coordinated their actions and if one partner was delayed the other waited for their buddy to arrive before pulling. Pairs of dogs, on the other hand, failed over and over again because they never pulled at the same time. It seems surprising that dogs couldn't master this, but, again, it makes sense if you think about the different social systems of the two species. Wolves in the wild hunt cooperatively, whereas free-ranging dogs individually scavenge for food and have a strict dominance hierarchy when it comes to sharing. Dogs may have failed to coordinate their actions, not because they didn't understand but because the subordinate dog in each pair was trying to avoid a fight.

In fact, when it comes to problem-solving, dogs and wolves have evolved completely different tactics. Wolves are much more persistent than dogs and, as Range tells me, 'If you're more persistent, then of course the chance that you'll succeed is higher, just by chance alone.' Dogs, on the other hand, exploit their relationships with people. If you set dogs and wolves an impossible problem, wolves will try brute strength as well as every trick in the book to succeed. Dogs initially get frustrated, but they then do something quite different – they look to their owner for help. It's called 'referential communication' and human infants do the same. It's an efficient strategy, taking up less time and energy than persistence, and it makes perfect sense for dogs' evolutionary history – over time, they started to rely less on their groupmates and more on their human partners.

Another study with the rope-pulling task probed this further. This time, the partner was a human, not another group member, so the question was whether this affected the animals' cooperative tendencies. Now, both wolves and dogs succeeded and both also successfully waited for a human partner to arrive before pulling, although there were some differences in the way the animals behaved towards people. Dogs looked at their human partners twice as often as the wolves did and tended to wait for their humans to make the first move, consistent with the idea that they viewed the person as dominant. Wolves were far less deferential to their human partner; they were more likely to initiate the movement than dogs and, anthropomorphically, you could say they were a bit more 'businesslike' about each trial – come in, do the work, get the treat, job done. For Range, this is a key difference: 'If I tell a dog "no" the dog accepts "no", but the wolf asks, "are you serious?"' She says that wolves cooperate with each other and humans at the 'same eye level', but this isn't the case for dogs whose cooperation with us is based on subservience.

The WSC's wolves can work with humans in other ways too. In a task where dogs and wolves could only get food by indicating an out-of-reach treat to a human, wolves used humans as cooperative partners just as well as dogs, with both species preferring to work with a cooperative person rather than a selfish person who kept the reward for herself. Like dogs, wolves were also sensitive to fairness – they reacted when a partner was rewarded but they were not for doing the same action, and they also reacted when they were both rewarded but their partner received a higher-quality reward. Wolves, like some other animals (more in Chapter 6), seem to have a mechanism for detecting unfair behaviour, which may be linked to their extremely cooperative nature.

Much of the problem-solving work conducted so far has been social; however, the few studies that have been conducted indicate that wolves are also better than dogs at solving physical problems. Taken as a whole, can we say that wolves are just a bit more intelligent than dogs? 'I think it's different kinds of intelligence,' says Range. 'There's a reason why we have dogs in our house and not wolves, and why dogs are very much more successful now than wolves ever were. I don't think you can say that one is smarter than the other, they're just different.'

Wolves are more persistent at problem-solving, and more tolerant and cooperative with each another, which fits with their natural ecology – they need their pack members to help bring down prey, defend the territory and rear offspring. Dogs, on the other hand, are friendlier towards people from an early age, more sensitive to what people are doing and they excel at tasks requiring cooperation with a person, which fits better with their status as a domesticated animal. It seems likely that the ancestral wolf species that gave rise to modern wolves and dogs already possessed sophisticated social and cognitive abilities; for wolves, these were channelled towards other wolves, while for dogs, these were enhanced and directed towards humans. In recent years, cutting-edge molecular analysis has also shed some light on the genetic basis for these differences. Dogs and wolves differ in a particular segment of gene, which corresponds with the gene that is altered in people with Williams-Beuren syndrome. People with this rare disorder are typically hyper-sociable, with characteristic facial features and some cognitive difficulties. The finding that wolves and dogs differ in this equivalent segment of the genome suggests a potential mechanism underpinning the change from wolves into friendly, people-loving dogs.

Unlike dogs, who have demonstrated deceptive behaviour towards people, there is no evidence that wolves try to deceive others. Mech agrees: 'I know of no outright cheating behaviour, but wolves do steal from each other.' Range describes how young wolves and pups will try to manipulate older animals with food into sharing with them, but notes that it's not the same as deception. 'It's more like a game that the whole pack plays – they don't need to be deceptive.'

Deception makes little sense for a wolf – it doesn't fit with their natural ecology and behaviour. So how should we be thinking about wolves? Range is pragmatic: 'They are a predator and they are potentially dangerous. And this is something that one should never forget. But on the other hand, wolves are so similar to how we were a long time ago. It would be very nice to see these parallels and maybe then there would be more respect for these animals.' For Mech, who founded the International Wolf Center in 1993 to promote objective science-based information about wolves, one of his goals is to encourage the public to think about wolves the way that they think about many other large predators. Not demonised, not romanticised, not vilified. No special status at all. Just an animal that is left to get on with surviving.

Fact or fiction?

It seems like Aesop got this one wrong. Yes, the wolf is an apex predator: it kills animals for food just like all other carnivores and in areas where it comes into conflict with human populations or contracts rabies it can pose a danger. But being an instinctive, wild carnivore isn't the same as being an evil villain, skulking in the shadows to plot our demise. We need to remember that we created these stories about the wolf by channelling our own fears and behaviours into them: we created the 'Big Bad Wolf' as the antithesis of all that is good in our world. It's the darkest form of anthropomorphism.

Wolves need a PR overhaul and for so many reasons. They are the closest living relatives of our favourite canine companions and they overlap in many of their behaviours; what's more, their intelligence, family loyalty and cooperation are traits that we value highly in our own society. Impressively, despite centuries of demonisation stacking the dice against them, they've managed to adapt and survive in a hostile world. And most tangibly, as an apex predator, their presence might help us to restore balance to ecosystems that are spiralling badly out of control. Yellowstone National Park provides the famous example. Devoid of top predators for decades, the park was saturated with elk and the reintroduction of wolves has undoubtedly played a role in bringing their population back under control.[*] It's impossible to know how big a contribution the return of wolves made to the recovery of Yellowstone (populations of other predators, including cougars and bears, have also increased) and many other factors are likely at play; nonetheless, what ecologists do agree on is the key role of apex predators in an ecosystem. It's not to say that we should be reintroducing wolves to every place that they once roamed: our footprint on the planet has expanded so much, that will never be appropriate. But let's judge wolf reintroduction efforts on the facts and the science, not on knee-jerk reactions taught by the stories of an ancient era.

We need to find a replacement for the wolf character, another animal to step inside the sheep's metaphorical clothing. And what that really means is identifying the animal that shows the best evidence for deception. Fortunately, there's a lot of it around.

---

[*] However, popular claims that this has led to a cascade of other effects, including the recovery of overgrazed willow and aspen, the return of beavers, healthier fish stocks and more, while making a wonderful story, are controversial.

In the 1980s, in southern Africa's dry savannah, psychologists Dick Byrne and Andy Whiten of the University of St Andrews, Scotland, were studying the effects of altitude on baboon social group size. In the dry season, food becomes scarce and troops must work harder to find food. While throwing out quadrats and measuring food availability, Byrne and Whiten's attention was drawn to the baboons' social behaviour. On one occasion, they watched an adult female called Mel digging in the cracked earth for deep onion-like tubers called corms. She had found one and was working to uncover it. Nearby, a juvenile male called Paul was watching carefully; no other baboons were in sight. Mel was nearly there when Paul let out an ear-splitting scream and his mother, Spats, rushed to the scene. Spats turned on Mel and chased her away, leaving Paul to claim Mel's prize.

Deception is one of those wonderfully broad, everyday terms, like 'imitation' or 'cooperation', that has long been used to cover all kinds of examples that fit with our intuitive understanding. It's all about communication, which biologists talk about as the transfer of information from 'sender(s)' to 'receiver(s)' to influence current or future behaviour. Human speech or wolf howls are obvious forms of communication, but information can be transferred in a multitude of other ways, tapping into the full sensory spectrum. That could be a plant releasing airborne chemicals to attract aphid-killing insects, the bright warning colouration of a cinnabar caterpillar, the dance moves of the blue manakin or the earth vibrations transmitted and detected by elephants who are many miles apart. Whatever modality is used, the sender is signalling something to another organism. The tricky part is that the information being signalled isn't always honest, like Paul's alarm call. The term 'deception' applies to the communication of dishonest information, which could be any form of cheating or trickery that has the effect of misrepresenting the world to another, irrespective of how it achieves that. Under this umbrella term, deception occurs across the animal kingdom and, while the term intuitively sounds as if it must be linked with intelligence, much simpler forms exist too.

At the least cognitive end, countless examples of the most exquisite mimicry demonstrate the extraordinary power of evolution by natural selection. This form of non-verbal communication usually functions to misrepresent reality to a member of a different species. It may be 'defensive', if the sender is some way down the food chain and wants to trick a predator into not spotting it. For example, there are caterpillars that are indistinguishable from bird droppings, butterflies that seem to disappear into the leaves upon which they rest, stick insects that could not look more stick-like and weevils that are not just the colour but texture of bark too.

Mimicry can also be 'aggressive' if the mimic is a predator and its mimicry fools prey. Larvae of the green lacewing, for example, show a 'wolf in sheep's clothing' strategy to live among their prey, the woolly alder aphid. Woolly aphids, as their name suggests, look like tiny sheep because they are covered in white 'wool' (in reality, waxy strands produced by the aphids for protection) and they are usually fiercely guarded by ants.* The lacewing larvae have taken the role of the wolf quite literally: they disguise themselves by stealing some of the woolly wax and covering their own bodies with

---

*Ants and aphids provide a well-documented example of a mutualistic relationship. Aphids feed by pumping large quantities of plant juices through their bodies, much to the consternation of gardeners who see the effects of these sap-suckers on their plants. The reason aphids suck so much out of the plants is to satisfy their requirements for nitrogen. However, this results in them taking in excess carbohydrate, which they do not need and instead excrete as a product called 'honeydew'. This sugary solution is an energy-rich food source prized by ants, who in turn fiercely protect their 'flock' from predation by other creatures.

it – and as a result they can walk straight past the ants and feast on the aphids. The lacewing larvae are manipulating the ants' visual and olfactory systems in order to misrepresent the world to them, an incredible evolved strategy for sneaking an easy meal.

Mimicry can be non-visual too. Bolas spiders hunt moths and, as the name suggests, these predators use a bolas (in this case, a single line of silk with a sticky drop of glue at the end) to capture their prey. When a moth approaches, the spider uses one leg to whirl the line around, trapping the moth when it's hit by the glue. Why the moth would be foolish enough to approach this obvious death trap comes down to a fantastic example of chemical mimicry. The bolas spider releases special scent compounds into the air that so closely match the species-specific sex pheromones released by female moths to signal mating, that male moths just can't say no. The irresistible lure of sex completely overwhelms the male moth's sensory system.

Mimicry in all its forms shows how weird and wonderful evolution can be – and aggressive mimicry certainly seems to fit with Aesop's sneaky wolf. But there's a fundamental difference. The rogue's gallery of animal mimics has been assembled by millions of years of adaptation. Much of it is hardwired into that animal's genetic make-up, meaning it is produced inflexibly, in a stereotypical and species-specific way. The lacewing larvae is showing behaviour at least, so that involves more information processing than the caterpillar that entered the world looking like a bird dropping, but its actions are nonetheless hardwired. This is categorically different to Aesop's wolf, which intentionally deceived its prey. We need to look a little harder to find a satisfactory replacement and that means thinking a bit more about what 'intentional deception' means.

Back to Paul the sneaky baboon. This wasn't Byrne and Whiten's only interesting observation. And yet, the only other

reports of similar behaviours at that time came from Jane
Goodall and the Gombe chimpanzees. Given that chimpanzees
are more closely related to humans than they are to baboons,
that seemed to make sense. 'We came back very impressed
with these baboons,' says Byrne, now emeritus professor at St
Andrews. 'And we discovered that most primatologists had
seen things like that, they just didn't dare publish them for fear
of being laughed at. We decided to take the risk.' To
differentiate their observations from simpler forms of
deception, Byrne and Whiten introduced the term 'tactical
deception'. The wording was deliberate – they proposed that
flexible deceptive acts, of the sort they had witnessed in the
baboons, could be considered as short-term 'tactics', in contrast
with inflexible, species-specific behavioural 'strategies' seen in
the lacewing or bolas spider. In essence: 'Tactical deception
occurs when an individual is able to use an "honest" act from
his normal repertoire in a different context to mislead familiar
individuals.'

Paul's crafty alarm call worked perfectly to alert Spats
because it was rare. Deception doesn't work if you pull the
same trick over and over – the boy who cried wolf learned
this the hard way. Deceiving your groupmates is also risky.
Get caught and you might get punished, either physically or
by exclusion. For example, rhesus macaques live in active,
noisy groups of tens or hundreds of animals. They forage on
their own and when they find something, they call to alert
the rest of the group. Except that they don't always call. One
study on Cayo Santiago, Puerto Rico found that troop
members were alerted less than 50 per cent of the time. Not
calling seems like an easy way to get ahead but it comes at a
cost. When cheats were discovered, they were frequently and
aggressively punished, and they ended up eating less overall
than honest individuals. The point is that if you're going to
deceive your groupmates, you need to be sure you won't get
caught and that means the behaviour needs to be both subtle

and perfectly timed. What this means for interested researchers is that it's really, frustratingly, difficult to study in wild groups.

To advance the topic further, Whiten and Byrne sent a questionnaire to 115 expert primatologists, asking them for records that matched their definition of deception. Fortunately, most of them kept meticulous records of all their observations, so numerous reports were shared. There was Dandy, a male chimpanzee who was trying to woo a female by exposing his erect penis at her,* but quickly covered it with his hands when an older, dominant male appeared. Or the female olive baboon who edged up to a male that was feasting on a freshly caught antelope. The female started to groom the male, who was known for his unwillingness to share, and when he 'lolled back under her attentions' she grabbed the carcass and ran. Byrne and Whiten pulled them all together and grouped them, publishing the resultant catalogue of tactical deception in 1988, together with a call to action for further research. It provoked an onslaught of commentary from psychologists, many of whom wrote strongly worded arguments about what they saw as an unjustified leap to invoking mentalistic explanations of the behaviour. 'The tactical deception work I was involved with in the 80s certainly showed "there's a lot of it about",' says Byrne, but he's quick to point out that not all these species are contenders for replacing Aesop's wolf. 'I guess we got a lot of publicity by using the word "tactical". But if you read the definition, it absolutely does not specify intentional. As carefully as possible, we avoided suggesting that tactical deception equated to intentional deception.'

At that time, the prevailing view among scientists was that intelligence had evolved to help individuals deal with challenges in their physical world; for example, how to find and access food, remember safe places or notice danger signals.

---

*This is normal behaviour for chimpanzees.

Not everyone was convinced, and from the mid-1960s there was increasing interest in the evolutionary link between social factors and intelligence. Notably, in 1966, based on her research with lemurs, Alison Jolly published an important paper proposing that primate social life, rather than tool use, 'provided the evolutionary context of primate intelligence'. Psychologist Nick Humphrey advanced the idea in 1976 with his influential essay, 'The social function of intellect'. In it, he argued that practical challenges in an organism's environment could not account for the evolution of advanced intelligence; instead proposing 'that the chief role of creative intellect is to hold society together'.

Byrne and Whiten pulled it all together, publishing their edited volume, *Machiavellian Intelligence* in 1988. The title was inspired by Niccolò Machiavelli, a diplomat and philosopher in the Renaissance who wrote *Il Principe* ('The Prince'), a book of advice for an aspiring prince on how he should act to achieve power and glory. They were not the first to link Machiavelli to ape behaviour: in *Chimpanzee Politics* (1982), Dutch primatologist Frans de Waal noted that much of what Machiavelli wrote could be applied to what he had observed to be complex, political chimpanzee groups. Byrne and Whiten's *Machiavellian Intelligence* hypothesis proposed a much more explicit link. They argued that the evolution of intelligence in primates had been driven by selection for social manipulative capabilities, because dealing with other individuals in a group was more challenging than dealing with technical problems. Species that tended to form larger, more complex social groups were expected to be under stronger selection pressures than those with simpler group structures or solitary species, resulting in differences in intelligence.

Machiavellianism, narcissism and psychopathy today make up the 'dark triad' in psychology, a suite of negative personality traits characterised by a lack of honesty and humility. It's therefore no wonder that when people hear about Machiavellian intelligence they automatically think about the more malevolent side of social behaviour. Yet, as Byrne has emphasised, this is a misunderstanding. The 'manipulation' produced by evolution is at the genetic level and while it may result in 'dark nastiness' between individuals, it can also take the form of cooperative, kindly behaviour. Just as Machiavelli advocated, the 'right' way to behave towards others depends on circumstance. That might be strategies to outcompete others or it might be behaviours that increase cooperation, as in wolf packs.

The evolution of cooperation has been a thorny issue for scientists since Darwin's time and we'll consider it in more detail in Chapter 6. Competition, on the other hand, has been a much easier concept for biologists to get on board with, particularly after Richard Dawkins' *Selfish Gene* (1976) focused attention on self-interest. Compared with helping, it's much easier to see how natural selection could favour the evolution of traits that put you ahead of others.

Today, we know that tactical deception (as originally defined) isn't restricted to primates: many examples exist and in a remarkable diversity of species. The distraction displays of lapwings, along with many other species of wader, are a good place to start. Lapwings, like other ground nesting birds, are highly susceptible to predation during the breeding season. With just one brood per year, it's crucial that parent birds can protect their vulnerable offspring. They have a couple of tactics to help. First, they may visit and guard decoy nests when potential predators are in the vicinity, helping to throw them off the scent. If the predator comes too close to the nest, another behaviour kicks in. The parent bird hops around in front of the predator, wing outstretched in a feign of injury, the behavioural equivalent of shouting, 'Easy meal over here, look at me!' It's a convincing act and often successful – the predator switches its attention to the parent bird and once it's moved far enough away from the chicks the parent simply abandons the lie and flies to safety. It's thought that this behaviour earned the lapwing its collective noun of a 'deceit'.[*]

There are several good examples of deception in food-storing species because caching is so susceptible to thievery. Why go to the effort of finding and storing your own food when you can just watch another and then steal it? In this case, both the cacher and the thief are trying to manipulate each other: the cacher will benefit if it can bury its food without being seen, while the thief will benefit if it can watch where the food is buried without being seen. Squirrels provide a classic example. If, when a cacher is burying its nuts, there is another squirrel in the vicinity, the cacher will return to dig them up and re-bury them in other sites, which likely acts to confuse the thief. They also trick the trickster by making fake burials. Nut in mouth, they go through the motions of digging a hole, seemingly pressing something into the ground and covering it over with leaves before bounding off. It often works, meaning that while the thief is digging up the empty store the cacher can get on with burying the real nut unwatched. A fascinating expansion of the squirrel work is that scientists working in artificial intelligence can now build robots that are capable of deceiving others, by plugging in rules based on real squirrel behaviour.

Producing fake alarm calls is another great example. It's been known since the 1980s that several birds do this and none better than the fork-tailed drongo. These are common birds in southern Africa and known for stealing food. They

---

[*]Indeed, 'plover' was used in the seventeenth century as a nickname for prostitutes and other 'deceitful women'.

follow foraging meerkats and other birds, such as sociable weavers and pied babblers, and capitalise on their foraging activities by either swooping down to catch flushed-out prey or bullying them into dropping it. It's not all bad for the foragers though – from their vantage points in the trees, the drongos can spot a range of predators and when they do they emit characteristic warning 'chinks'. This means the foragers can spend more time looking for food and less time being vigilant. On the surface, it seems like a mutually beneficial arrangement; look a bit closer though and it's clear that the drongo has the upper hand. When a babbler or meerkat finds a large food item, the drongo often calls dishonestly, and research has shown that babbler and meerkat targets flee on 82 per cent of these occasions. It's an easy way to get a free meal. If the forager ignores the drongo, it does something even more impressive: it changes its call. In total, drongos have been reported producing up to 51 different warning calls (each individual's repertoire is between nine and 32), 45 of which are mimicked alarm calls of other species. As discussed earlier, deception only works when it is not detected, and if the drongos 'cried wolf' with the same call too often they would quickly be ignored. By mixing things up they stay one step ahead of their unfortunate Kalahari companions.[*]

Sex is the other main driver for tactical deception. Male topi antelopes, for example, produce fake alarm 'snorts' to increase mating opportunities. During the breeding season, males compete to establish territories and nonterritorial females, who are only in season for roughly one day of the year, visit several of these. Unlike many other ungulate species, the males don't fight each other to demonstrate their qualities to the females; instead, they trick the females.

---

[*] But before you vilify the fork-tailed drongo too much, know that they are currently the only known host species of the African cuckoo, itself a frightful cheat. In nature, the player also gets played.

Researchers found that when a female attempted to leave a male's territory, the male frequently produced a fake alarm call and exhibited classic anti-predator behaviour, stopping and gazing to a point in the distance. It worked – most females turned from the direction of the 'threat' and walked back into the male's territory, which often resulted in additional matings with the cheat. At present, females seem unable to distinguish between true and false alarm snorts, and since the cost of getting it wrong and ignoring a genuine alarm snort is more severe than responding to all snorts, in this evolutionary arms race the males currently have the upper hand.

In Sydney harbour, male mourning cuttlefish have taken things one step further. Cuttlefish are known for their ability to change colour and pattern; they do so to camouflage themselves against their surroundings and they also display striking patterns to females during courtship. What's really remarkable is that they can be deceptive in their colour change. When a male is trying to impress a female, he displays a brilliant pulsating zebra stripe pattern. The female has a much duller brown and cream mottled pattern. Researchers were watching trios of cuttlefish, where a focal male was already displaying towards a female and another male then appeared. They observed something quite astonishing. The focal male made sure to be positioned between the female and his rival, and he then split his display down the middle of his body. On the half of his body that faced the female he continued with pulsating stripes; on the other half, the one that faced the rival male, he switched to the mottled female pattern. The researchers suggested that males were reducing competitive interactions with the other male by tricking him into responding to the female mimic and not the true female. It seemed to pay off: some sneaky males were seen to transfer sperm to females. What's more, cuttlefish were flexible about when they employed the tactic: they didn't do it when there was more than one male in the vicinity, possibly because it would be too difficult to orient effectively to trick multiple males.

There are many more examples, all consistent with the criteria of an act from the animal's honest repertoire being used in a different context to deceive another individual. And while it should be clear that examples of deceptive acts are not uncommon in the animal world, this doesn't mean that they equate to human deception. 'For most mammals that I'm aware of, I don't think there's a shred of evidence that they really understand mental mechanisms,' Byrne explains. 'They use fascinating things, they learn an enormous amount about the social world, but it doesn't make them like people.' He's keen to emphasise that this doesn't mean they lack intelligence – being able to learn a subtle deceptive tactic from one encounter in a complex social world requires brainpower, and that would have resulted in selection for very rapid learning and efficient memory. 'I would still say a lot of these animals are intelligent. What they achieve is pretty clever, even if they don't understand the situation in the intentional way we do.'

Using the reports compiled in their catalogue of primate tactical deception – and controlling for observer effort – Byrne and colleagues showed that the size of a region called the neocortex positively predicted the extent to which deception had been reported. Few reports of deception were made in the lemurs and bush babies, which also had the smallest neocortex. The most reports were made in the great apes, which also had the largest neocortex. This supports a learning-based explanation for deception, because considerable brain capacity is thought to be needed to allow the enhanced memory and learning abilities considered crucial for animals to remember details of their groupmates and the best ways to manipulate them.

Learning and memory are clearly essential components of manipulative behaviour, in us as well as other animals; where scientists disagree is whether they can explain it all. That's because some examples of tactical deception don't seem to be accounted for by learning alone and that raises the possibility of a more cognitively advanced explanation. This was the case for several of the examples Byrne and Whiten had catalogued, leading them to publish an article in 1992 in which they concluded that great apes, but not monkeys, had some understanding of others' intentions.

To intentionally deceive another, you need to have a good idea of how they are going to behave in a situation and that leads us to something called 'theory of mind'. This is the understanding that other individuals have thoughts, wants, beliefs and feelings that may differ to our own. It is at the heart of our day-to-day social interactions, enabling us to make predictions, judgements and inferences about why other people are doing certain things or how they will respond to things that we say or do. Theory of mind develops through early childhood and is thought to be fully formed by the age of about four years. The most challenging stage is being able to attribute 'false beliefs' to others. That's the understanding that not only can other people have different thoughts and beliefs to you, but that these might not reflect reality. The classic experiment to test for false beliefs is the 'Sally Anne' task, which takes the form of a little puppet show. Children watch while researchers act out a story and they are then asked questions. For example, Sally is given a chocolate bar and to keep it safe she places it inside her basket. She then leaves the room and the children watch as Anne steals the chocolate from Sally's basket, placing it inside her box. Sally then comes back into the room and the child is asked: 'Where will Sally look for her chocolate bar?' At the age of four years, children point to Sally's basket, because they understand that what Sally thinks does not reflect reality. At three years of age, however, children tend to incorrectly point to Anne's box, suggesting that they haven't yet grasped that Sally's understanding of the situation is different to their own.

A fully–fledged understanding of the mental lives of others is central to what makes us human. It's why we can forgive people for saying stupid things before they know the full facts of a situation; why we can teach and counsel effectively. It's also what allows us to be master manipulators. Grifting, hustling, cheating – however you think about it, it's underpinned by our exceptional ability to manipulate other people's beliefs.

In the past 40 years, numerous studies have been conducted with different species to address what has become something of a holy grail in the field of animal cognition – do other animals also have a theory of mind? Ironically, the term 'theory of mind' was introduced in 1978 by primatologists working on chimpanzee behaviour, but in subsequent decades it became thought of as a uniquely human behaviour resulting from language and culture. Although Whiten and Byrne's synthesis of observational data suggested that great apes may intentionally deceive, experiments with captive individuals concluded that there were significant differences in the ways that humans and animals interpreted the social world. For example, chimpanzees begging for food from a human experimenter showed no preference between a person with a bucket over her head and one with a bucket next to her head, indicating that they had very limited understanding of what the person could see and know. There were many other experiments of a similar nature and in 1997, when psychologists Mike Tomasello and Josep Call published their synthesis of primate cognition (called, funnily enough, *Primate Cognition*), they debunked the idea, concluding that there was no clear evidence that non–human primates could attribute mental states to others. The consensus was that theory of mind reflected a genuine cognitive difference between humans and other animals.

Science doesn't stand still, and in an important publication in 2000, evolutionary anthropologist Brian Hare and colleagues argued that the experiments to date were not truly representative of natural social interactions. They noted that chimpanzees were far more likely to be motivated by competitive, rather than cooperative, social interactions, and supported this with research showing that a subordinate chimpanzee was more likely to go for food that a dominant animal could not see. It kick-started a new line of investigation and by 2008, some 10 years after their book had been published, Call and Tomasello concluded that things had changed. A flurry of new experiments in which apes competed either with another ape or a human provided evidence more consistent with theory of mind, indicating that the animals took advantage of what their rival could and couldn't see, hear or know. They also seemed to be sensitive to the goals and intentions of others, for example by discriminating between people who accidentally and intentionally failed to cooperate with them (by either spilling or deliberately pouring away a tasty fruit drink) and even completing the failed actions of others.

Theory of mind research has not been confined to apes. In monkeys, an impressive body of research led by Laurie Santos, professor of psychology at Yale University in Connecticut, and team on Cayo Santiago have provided intriguing evidence for the ability. Cayo Santiago is a special place for primate research. The brainchild of pioneering primatologist Clarence Ray Carpenter, in 1938 he shipped several hundred rhesus macaques from India to the island and today more than 1,000 of these Old World monkeys live here; an unusual outpost to their otherwise Asian distribution. It has earned itself the nickname 'Monkey Island' and is the longest-running primate field site in the world, providing valuable genetic, population and behavioural data. The monkeys are completely habituated to people and try to steal food from them at every opportunity, which allows Santos to test predictions relating to theory of mind using similar competitive scenarios as have been used with apes. For example, they've found that macaques are much less likely to steal from someone who can see them compared with someone who can't. This preference was shown for someone whose back was turned over someone who was facing the monkeys, but it was maintained in much more subtle conditions too, such as when the only difference between the people was that one held a little barrier over their eyes and the other held the same over their mouth. The overall picture from her many studies is that rhesus macaques have some understanding of what others can see, hear and even know, i.e. components of a theory of mind.

Outside of primates, we have already looked at how squirrels employ strategies to protect their caches from thieves and several species of corvid do the same. Studies of western scrub jays and ravens suggest that their cache-protection strategies may be underpinned by a theory of mind. The scrub jay research has been headed up by Nicky Clayton and the raven work by Thomas Bugnyar, both of whom we met in the previous chapter. One of Clayton's intriguing findings is that scrub jays that are caching when watched by others will later dig up and rebury the food, but only if they have personally experienced being a thief. Jays that had never raided other birds' caches were less likely to recache their own food, even when another bird was watching, raising the possibility that the birds project their own experience of pilfering on to other birds to predict how they will behave.

Bugnyar has been fascinated by theory of mind in ravens since his studies began. After observing how ravens adjusted their food-caching behaviour in the presence of other individuals, he conducted studies to look for evidence of tactical deception in the birds. 'It's really all about seeing,' he tells me. 'The one that is caching should make sure not to be seen, whereas the other needs to watch but ideally without being seen, otherwise the first won't cache the food. This was the whole starting point.' The question then became, what do ravens understand when they are engaged in such behaviour? Over the years, Bugnyar and colleagues have tested the competing hypotheses one by one, finding that the ravens' behaviour was consistent with a theory of mind explanation, but unable to demonstrate it conclusively. Until they conducted the 'peephole' study.

The study capitalised on ravens' natural anti-thievery tactics, but with a very clever twist. A raven was brought into a testing room and provided with food that it could hide around the room. Initially, a window in one wall was uncovered, allowing the bird to see another raven in the next room. In these trials, the test raven behaved as it would normally in the presence of another bird, stashing food away from the window and behind barriers that occluded the other bird's view. In subsequent trials, the window was covered over but raven calls were played through a speaker, creating the impression of another bird in the neighbouring room. The ingenious part of this experiment was that there were also little peepholes in the testing room's walls. Every test subject had been able to learn about these during pre-testing trials, in which they were allowed to peer through the peepholes into the testing room. During testing, if the peepholes were closed, the raven cached as if it was not being watched. But if a peephole was open, the bird behaved as if the window was open and another bird was watching. 'It was so cool on that first trial,' says Bugnyar. 'All of them did it, it was so clear.' It suggests that the bird was using its own experience of looking through the peephole to predict what the 'snooper' could see. To date, it stands as the best evidence that birds possess a theory of mind.

At this point, there are several contenders to take the role of Aesop's wolf, but one question remains: how do they fare on the false belief task? Up until recently, this seemed to be the stumbling block. Attempts had been made to replicate the Sally Anne task with apes, using videos created for human infants in which a woman and a little puppet slowly and calmly moved an object between boxes. Apes will happily sit and watch videos, but these didn't work well.

'They find it completely boring!' says Chris Krupenye, postdoctoral researcher at Durham University and Johns Hopkins University. 'They don't pay any attention.' Krupenye teamed up with Fumihiro Kano, assistant professor at the University of Kyoto, to develop an adjusted experimental approach for studying ape theory of mind. Together with their collaborators – Tomasello, Call and Professor Satoshi Hirata – they created new videos that were much more engaging for the apes; while still based around the Sally Anne task, they were spiced up with conflict, aggression and a weirdly enthralling 'King Kong' character (in reality, Hirata, who agreed to dress in a full-body gorilla costume for the experiments).

The basic story was that an actor was about to stash his stone inside one of two boxes when Kong rushed in and stole it out of his hands. Crouched in front of the boxes, Kong appeared to perform a ritualistic victory dance, banging the stone several times against the chamber's walls (Krupenye explains that this was to emphasise the ongoing conflict), before placing it inside one of two boxes. Kong then threatened the actor into leaving the room and he then moved the stone to the other box. The actor re-entered the room and approached the boxes – but in which would he look? Using an innovative eye-tracking set-up developed by Kano, the team could track the ape's gaze direction before the actor's actual search and they used that as a proxy for where the ape expected the actor to look. It revealed that, just like four-year-olds in the Sally Anne test, the apes overwhelmingly looked towards the first box, where Kong had originally stashed the stone in sight of the actor.

This on its own did not provide strong enough evidence for ape theory of mind – sceptics pointed out that the same results would be seen if the apes simply looked towards the box in which the actor had last seen the stone, which isn't the same as reading their minds. Disentangling the two possibilities is incredibly difficult, which is why there is still so much debate about theory of mind in non-human animals. So, the team followed up with another study, which was published in 2019. Just like the earlier test, Kong played the role of villain,[*] scampering in and stealing the actor's object, placing it into a box (called the 'target') and then moving it again. But this time, when Kong threatened the actor he didn't leave the room, he just stepped back behind a screen. The screen was white and its edges were decorated with sparkly tinsel to make them more visible to the apes. It was large enough to occlude the actor's head and torso, but not his legs and feet. While the actor stood behind it, Kong took the object from the target box and placed it into the second 'distractor' box, and then he removed it altogether and scuttled out of the room. The actor then returned to the apparatus and reached for a box, and the team again used eye-tracking software to infer the apes' predictions. The crucial difference between this and the previous test was that before the apes were tested, they had the opportunity to learn about the screen's properties. Half of the apes experienced an opaque screen, behind which objects disappeared. The other half experienced an identical looking 'trick' screen, which was made of mesh and behind which objects were still visible. The question was whether the apes used their own experience of the screen to predict what the actor could see and therefore know about Kong's removal of the stone. The team found

---

[*] Kong had to be the bad guy – the team did try and switch the roles in one set of trials, but the apes were apparently so obsessed with looking at him that they ended up ignoring what else was going on.

that apes who had experienced the opaque barrier showed a higher looking score towards the target box, which is consistent with them expecting the actor to search for the object where it was before they went behind the barrier. In contrast, the apes who experienced the trick screen appeared to have no expectation that the actor would look in either of the boxes, which is what you'd expect if they thought the actor had already seen the object being removed. These results are consistent with the prediction that the apes were attributing to the actor either a false or true belief, depending on what they thought the actor could see. It is a groundbreaking study and provides the best evidence yet for theory of mind in non-human animals.

Santos and colleagues have also tested rhesus macaques on adapted versions of the Sally Anne task and have not found any evidence that the monkeys can attribute false beliefs, which is consistent with the idea of a genuine cognitive difference between monkeys and apes. In ravens, it's yet to be seen – would it even be possible to design a raven version of the task? Bugnyar pauses. 'The last chimp studies are so cool. Really impressive ... But I'm not yet fully convinced about the right measures with the ravens. It's so easy to do a test, but if you don't set it up right and have the proper controls it can't tell you anything.' He explains that once the birds have been used in one test, they can't be used in a related experiment because it's impossible to exclude the effect of learning on their first-trial behaviour.[*] Nonetheless, it is a key area of interest for Bugnyar, so stay tuned for the next development to this story.

---

[*] This is a persistent challenge for cognition researchers working with a limited group of animals, which is the norm for those working with large mammals, birds or reptiles, who cannot simply go and collect a few more animals.

Humans are the masters of mental manipulation, but many other animals 'cry wolf', albeit in simpler, non-intentional ways. At present, the convergence of evidence across different studies for ravens, rhesus macaques and apes suggests that they have a kind of theory of mind, although so far only apes have shown that they may be able to attribute false beliefs. There's still controversy, of course, as Krupenye points out: 'It's hard to identify a method that everyone agrees can provide consistent evidence for or against theory of mind.' And even if it were, there's plenty more work to be done, both to work out the finer details and to tease apart alternative explanations. Nonetheless, apes' apparent sensitivity to false beliefs, combined with the fact that they can capitalise on others' ignorance, suggests that they may have the cognitive capacity for intentional deception. As yet, there's no evidence that apes can implant false beliefs in order to deceive another. Nonetheless, in our quest to find a wolf replacement, 'The Ape in Sheep's Clothing' would be a little closer to biological fact.

CHAPTER THREE

The Dog and its Shadow

It happened that a Dog had gotten a piece of meat and was carrying it home in his mouth to eat it in peace. Now on his way home he had to cross a plank lying across a running brook. As he crossed the brook, he looked down and saw his reflection in the water beneath. Thinking it was another dog with another piece of meat, he made up his mind to have that also. He made a snap at the reflection in the water, but as he opened his mouth the piece of meat fell out, dropped into the water and was never seen again.

The snow doesn't bother Jethro. A German Shepherd-Rottweiler cross, he hurries down the trail, nose skimming over flattened ice and drifted powder. It is just after dawn, and Boulder Creek in Colorado is an artist's palette of cool blues and greys, the weak winter sun still too low to banish the night. Left, right, back again, Jethro is immersed in his world of sights, sounds and smells. Suddenly, he swings off into the vegetation and as he explores his owner does something very odd. He has stopped by a fresh patch of 'yellow snow' and, with gloved hands, he deftly scoops it up, transports it down the trail and then packs it neatly back into the snow. Jethro, once content with sniffing in the bushes, races to join his owner and then notices the urine-soaked snow. Another dog! He sniffs at the patch for several seconds, gleaning what information he can about the owner's identity and then, curiosity satisfied, he urinates over the top.

Jethro's owner, Marc Bekoff, conducted this strange ritual for five winters between 1995 and 2000. He made a mental note whenever he saw Jethro or another dog urinating into the snow and then, while Jethro was preoccupied, he transported it further up the trail and recorded how his dog reacted upon its discovery. He found that when Jethro encountered snow soaked in other dogs' urine, he spent time sniffing at it and frequently urinated over it; in contrast, when he encountered his own he paid far less attention. Bekoff, now professor emeritus at the University of Boulder, Colorado, concluded that Jethro recognised the latter was his own, writing that: 'Animals such as Jethro ... possess some sense of self or ownership over their body and smells.'

It's the sort of research project that ends up shortlisted for the Ig Nobel Prize, but Bekoff had a serious motive. He was trying to open a door into Jethro's world, to learn whether Jethro recognised himself and knew that he was a separate entity to other dogs, humans and his environment. In other words, he wanted to know if Jethro had a concept of Jethro. The standard approach to assessing self-recognition was with a mirror and this is where it relates to Aesop's dog, who saw only another dog in his reflected image. But Bekoff was not convinced by this experiment, so he had set out to devise something a bit more 'dog-relevant'.

★ ★ ★

On 28 March 1838, a year after he returned from his voyages on HMS *Beagle*, Charles Darwin visited London Zoo. At this time his thoughts about evolution were not yet fully formed; it would be 20 more years before publication of *The Origin of Species*. At the zoo, Darwin observed with great interest the behaviour of Jenny and Tommy, a pair of young orangutans who were dressed as humans and taught to do things like drink tea. Darwin's attention was drawn to the apparent similarities in the apes' emotional states with those of humans, leading him to write about their jealousy, excitement and fear, as well as their understanding of situations. Among the tests that he set the pair, Darwin placed a mirror into their enclosure, recording the following in his notes:

> At first they gazed at their own images with the most steady surprise, and often changed their point of view. They then approached close and protruded their lips towards the image, as if to kiss it, in exactly the same manner as they had previously done towards each other, when first placed, a few days before, in the same room. They next made all sorts of grimaces, and put themselves in various attitudes before the mirror; they pressed and rubbed the surface; they placed their hands at different distances behind it; looked behind it; and finally seemed almost frightened, started a little, became cross, and refused to look any longer.

These observations played a crucial role in Darwin's development of ideas about the evolutionary relationships between humans and apes. Darwin did not believe that animals were self-conscious in the same way as humans – for example, in thinking about where they came from or their own mortality. Yet, he also didn't believe that animals had no sense of self, just that it was much simpler than our own.

Darwin's notes of Jenny and Tommy do not suggest any evidence for self-recognition, yet today orangutans are one of the very small number of species that appear to recognise their mirror-reflected image. It's a contentious topic, and to fully understand why and how it relates to Aesop's dog, we need to go back to its beginning: to the late 1960s when a young graduate student named Gordon Gallup Jr. was shaving in front of his bathroom mirror. Gallup was thinking about potential research projects when the idea struck. 'It just occurred to me, wouldn't it be interesting to see if animals could recognise themselves in mirrors?' It wasn't until Gallup took his first academic job, over three years later, at the University of Tulane in Louisiana, that he revisited the idea. The Tulane National Primate Research Center was happy to host the project, allowing Gallup to place mirrors outside the cages of several pre-adolescent chimpanzees for up to eight hours a day. Initial responses included aggressive vocalisations, sexual and social displays, which are normal chimpanzee behaviours to another animal and consistent with Darwin's observations. But Gallup didn't stop there. In the next phase of testing he moved the mirror much closer to the cage and he started to witness a shift in the animals' behaviour. They began to inspect body parts that were only visible with the mirror, including the inside of their mouths and their genitalia; they picked food from their teeth and mucous from their eyes; and they also made faces and blew bubbles. 'At that point I was pretty much convinced that the chimpanzees had learned that their behaviour was the source of the behaviour being depicted in the mirror.'

'The problem,' Gallup adds, 'was that I didn't think many of my colleagues would be terribly enamoured with my subjective impressions of what had transpired.' He knew he needed a more scientific method to formally test self-recognition in the chimps, so he devised the 'mark test'. Each animal was anaesthetised, and Gallup used a Q-tip to apply an odourless red dye to the upper left eyebrow and the right ear, places that were not visible to the animal without the mirror, and far enough away from the nose or mouth to be sensed by another means. Once the chimp woke up and started eating and drinking normally, it was watched for a short period to see how many times it touched the newly dyed areas. The mirror was then returned to its position just outside the cage, and Gallup returned to his spy-hole. He found that without the mirror, just one of the animals touched a marked area. After the mirror was put back into the room, all of the chimps repeatedly touched their red eyebrow and ear. After touching the mark, some animals also peered expectantly at their fingers and sniffed them. Other chimps, who didn't get the opportunity to learn about mirrors before having the same red marks applied, showed no self-directed behaviour, suggesting that the capacity for self-recognition had been learned by the original subjects.

The significance of the experiment came for Gallup when he tested the same protocol with macaques. Even after hundreds of hours of mirror exposure none of the monkeys showed any evidence of mark-directed behaviour. He was most convinced by the behaviour of two hand-reared infant monkeys, which he provided with 12–14 hours of mirror exposure daily, for years. Even after a lifetime of exposure, the monkeys never showed evidence for mirror self-recognition. This was key, as it suggested a difference in cognitive ability between chimpanzees and monkeys: the latter just didn't seem to 'get it'. 'In a way,' reflects Gallup, 'I thought that was one of the most interesting findings. What the chimpanzees did was pretty much what I expected'.

Gallup published his pioneering study in 1970 and his mirror self-recognition (MSR) test has subsequently been rolled out to a diverse array of species. Fifty years later, it's still the gold standard test for self-recognition, and there are still very few species that convincingly 'pass'. For some researchers, including Gallup, only humans, chimpanzees and orangutans show conclusive evidence. It's a contentious stance, as others have reported positive results with several other species, including dolphins, orcas, one Asian elephant, magpies, monkeys, manta rays, little fish called cleaner wrasse and, most recently, horses. What's more, numerous species have failed, with notably mixed results for gorillas. If MSR is confined to humans and our closest relatives, it would be puzzling if gorillas entirely lacked the ability. In fact, the experiment seems to tap into a potentially confounding aspect of gorilla social behaviour – prolonged eye contact is considered an aggressive signal, so the animals may simply avoid looking at the reflection for too long. Other studies have found positive evidence, but only in gorillas that have been brought up in an 'enriched' environment with extensive human interaction.

Additional controversy stems from the supposed wider implications: that passing Gallup's MSR test reveals a capacity for self-awareness, which is the ability to become the object of one's own attention. Being self-aware, Gallup explains, may be a prerequisite for that elusive ability we encountered in the previous chapter: theory of mind. That's because self-awareness likely forms a foundational layer, providing the necessary cognitive architecture to be aware of others. And because of that, the implications of not demonstrating self-awareness have broader impacts; for Gallup, animals that fail the MSR test lack self-awareness and therefore cannot possess a theory of mind.

Not all researchers agree on Gallup's assertion that MSR is a proxy for self-awareness. In fact, it's fair to say that Gallup's mirror test is as controversial now as when it was first published. Psychologist Celia Heyes, for example, pulled no punches when she wrote in 1995 that not only did the test not provide evidence for self-awareness, but that the resulting speculations about self-recognition had 'led to a considerable waste of time and resources'. For Heyes and other sceptics, apparent MSR could be accounted for without any form of advanced cognitive ability. A chimpanzee that touched the red spot on its face while looking into the mirror could simply have learned to coordinate what it sees with what it feels. It's a subtle, but significant difference – an explanation that could theoretically allow for the animal to respond appropriately to its reflected image, but without any mental concept of 'self'. It is true that some animals (such as pigeons) can be trained to exhibit apparent mark-directed behaviour, yet as Frans de Waal has commented, when thousands of rewarded trials are required to shape the behaviour the responses are 'as meaningful as would be the literary talent of a monkey taught to type "to be or not to be".' What's more, Gallup's monkeys had ample opportunity to learn and never showed MSR – if simple learning processes could account for the behaviour, then surely it should be more widespread?

As the number and diversity of species that are reported to pass the mark test rises, the choice becomes either accepting that more animals have a sense of self than previously imagined, or that the test is not black and white in what it reveals. De Waal, for example, has cautioned against what he called the 'Big Bang' theory of self-awareness, in which species either have it or do not. Instead, he has suggested thinking about self-awareness more like an onion, built in layers, with different species possessing different gradations of the ability. It is certainly the case that the behaviours observed in chimpanzees towards their reflection are very different to the behaviours observed in the single Asian elephant that was said to pass, or the cleaner wrasse. A single test may not be able to reveal these different levels, meaning that more tests are necessary, including non-visual tasks.

Bekoff and his advisor, Michael Fox, tried to replicate Gallup's test with captive wolves, coyotes and dogs in the 1980s and the animals showed no interest whatsoever.* But they weren't convinced by the interpretation. Having spent decades working with dogs and observing their behaviour, Bekoff and other canine ethologists had major objections with the appropriateness of the test. They agreed that dogs showed no evidence for MSR, but dismissed the claim that this proved a lack of self-awareness; an absence of evidence, in this case, not necessarily demonstrating the absence of the ability. At the heart of the objections was the fact that dogs live in a strongly smell-based world, recognising other animals with their nose, not their eyes. A reflected image just doesn't provide the same kind of information, so the mark test wasn't a relevant tool to answer questions of self-recognition. In order to understand what dogs know about themselves, Bekoff argues, we need to change our whole way of thinking about this experiment. Instead of asking if dogs behave like we would in the same situation, we should put ourselves in their shoes and ask: what is it like to live as a dog and how can we use this information to investigate its sense of self?

* * *

---

*Unfortunately, despite submitting their findings to several journals at the time, they were never accepted for publication. This is a common and serious problem in scientific publishing, resulting from a mistaken belief that nothing can be learned from negative results. Of course, finding out that an animal doesn't do something is just as important for advancing knowledge of a field. Bekoff told me that several people had been in touch over the years saying they wish the results had been published because they then wouldn't have spent their time trying to run the same study.

Sitting with friends in the beer garden of our local pub one balmy summer evening, conversation turned to what we had wanted to be as children. Among the resulting childhood ambitions of architect, soldier and astronaut were two wonderful revelations: one friend's boyhood desire to become a train and my recollection that as a very young girl I had gone through a period of wanting to be a dog.*

How does one become a dog? For me, I'd dressed myself in brown and grey clothes, fastened long socks to droop down over my ears and proceeded to walk around on all fours, to the bafflement of my family. Surely this gave me some insight into what it's like? 'Walking on all fours is a start, at least!' agrees Alexandra Horowitz, head of the Dog Cognition Lab at Barnard College, New York. 'I don't know that we *can* ever know what it is really like to be a dog – but neither can we really know what it's like to be another person, for that matter.' Horowitz, who was mentored by Bekoff as a graduate student, is fascinated by how dogs integrate and make sense of information from the world around them. One of the concepts that she finds particularly relevant for interpreting behaviour is the 'umwelt' (pronounced 'oom–velt'), an idea proposed by German biologist Jakob von Uexküll in the early twentieth century, which roughly translates as an organism's 'environment-world'. Von Uexküll's theory was based on the notion that every animal perceives and experiences its environment in a different way, so that only specific bits of it are relevant for each. He illustrated this with the umwelt of a tick in a grassy field. Blind and deaf, these little animals use light-sensitive cells in their legs to help them climb to the top of a blade of grass. Subsequently, they use their sense of smell to detect the odour of butyric acid (which emanates from the

---

*Coincidentally, we were the only ones in the group to have grown up in Wales. I like to think that we were encouraged to think outside the box.

sweat glands of all mammals) and, once detected, they fall from the grass blindly. A temperature sensor confirms whether it has landed on something warm and if so, the tick simply needs to find the least hairy patch of skin to embed itself in.

For the tick, only certain stimuli within the overall environment of the grassy field are important: light, smell, temperature. The rest of the environment is irrelevant – the tick does not pay attention to the song of the skylark or the colour of the sky. The environmental cues that are important to a tick bear no resemblance to those of a kestrel, hovering over the same patch of grass, or a vole, crouched motionless at the base of the stems.

A key component of von Uexküll's theory and Horowitz's research is that we don't know what it is to be in the umwelt of another animal. Of course not. We regularly try (and often fail) to put ourselves in the shoes of another person, so can we truly believe that we know what it is to be a dog, crow or snake? What we can do is make predictions based on what we know about sensory abilities, says Horowitz. 'With a good understanding about what the perceptual and cognitive experience is of a dog – or any animal – we can take an imaginative leap into what it's like to be them … That involves, with a dog, understanding what it's like to smell one's way through the world.' This is where Bekoff's yellow snow test comes in. Dogs are governed by their noses, not their eyes, so if we want to think like them, we need to step inside this richly scented world.

★ ★ ★

When we talk about a 'smell' we are really just talking about our brain's interpretation and classification of particular chemical molecules. The smell itself is literally all in our heads. Imagine that you are walking past a bakery and they've just removed a batch of bread from the oven. As you inhale, air rushes up your nose into the nasal cavity, passing over a patch of mucus-coated sensory tissue called the olfactory epithelium. This tissue is densely packed with olfactory receptors, each one at the end of a neuron that connects with the olfactory bulb of the brain. As the chemical molecules contained in the inhaled air dissolve in the mucus, they cause specific receptors to fire, transmitting signals about the odour to the olfactory bulb. If the smell is familiar, the brain identifies it. If, for example, a molecule of β-phenylethyl alcohol reaches the olfactory epithelium of your nose, you will likely recognise it as rose. A molecule of amyl acetate is identified as banana. The smell of freshly baked bread is a more complex mix of several compounds, although 2-acetyl-1-pyrroline is the most significant in the smell of wheat crust.

Our sense of smell does all that we need, enabling us to detect rotten food, things that are burning or bodily odours. It is not our most important sense and that's because our world is predominantly interpreted through our eyes. We notice smells when they catch our attention for being attractive, repulsive or unusual, but for much of the day we are largely oblivious. Yet, if you constructed a map of three-dimensional space that showed all the airborne chemicals in gradients of strength you would see a staggeringly complex and shifting mosaic. No place would be devoid of scent. It might seem like that to us, but we just aren't built to detect them.

Our olfactory epithelium covers an area of about 10cm^2, which seems like a reasonable size for a patch of tissue up the nose, until you consider that a dog's, as a result of massive folding, can cover up to 170cm^2! And not only is it gigantic compared with ours, a dog's olfactory epithelium is also far more densely packed with nerve cells: we have in the region of 6 million olfactory receptors, while a dog has about 220 million or, in the case of a bloodhound, more like 300 million.

Why are dogs so phenomenally good at detecting smells? Well, they're mammals for a start and in general mammals are characterised by their outstanding sense of smell. It comes down to those olfactory receptors, the sensory cells in the nose that transmit chemical signals to the brain. In mammals, the genes responsible for olfactory receptors (OR) make up the largest multigene family. Dogs have in the region of 900–1,000 functional OR genes, and an additional 300 or so non-functional OR 'pseudogenes'. The only mammals that don't share the ancestral mammalian sense of smell are the ones that have secondarily lost it. The bottlenose dolphin, for example, has just 58 functional OR genes, about the same as a fruit fly. That's because when the long-distant ancestors of whales and dolphins took to the water, they began relying less and less on their sense of smell; there was no selective pressure to maintain the areas of the brain associated with olfaction. Primates are the other group that came to rely less on smell than sight, as we'll learn. We have about 400 functional and 400 non-functional OR genes, and although most people can recognise a broad range of odours, the absence of a receptor type can result in an inability to recognise a specific smell. For example, one in 1,000 people can't detect the smell of butyl mercaptan, which is the stinky compound released by skunks; one in 10 of us can't detect the almond-like smell of the highly poisonous gas, hydrogen cyanide; and it's thought that as many as 60 per cent of us are unable to detect the pungent odour of metabolised asparagus in urine.

Anatomically, there are some differences too. When dogs breathe in, most of the air sweeps down to the lungs, but some of it is diverted to the back of the nose where it passes over a specialised area of bony structures called turbinates, which are densely packed with olfactory receptors. When dogs exhale, the air isn't simply forced back out the way it came in, because this also forces out any incoming odours. Instead, spent air exits through little slits at the side of the nose, meaning that a dog can keep processing a smell while it is breathing. There's variation between breeds of course, with those bred for sniffer purposes far better than those bred as companions. Bloodhounds don't just have fantastic sensory systems: their loose, wrinkled facial skin helps to trap odour particles and those long, droopy ears trail along the ground, helping to sweep up scent and funnel it up to the nose. There's a reason why bloodhound tracking evidence is admissible in some US courts. Dogs can also wiggle their nostrils independently, helping them to pinpoint the direction of a smell and there's evidence that they use their nostrils differentially too: the right for novel and potentially threatening smells, the left for familiar, non-aversive smells (such as food) and scent from other species (such as human sweat for tracking purposes).

This all adds up to dogs having a sense of smell that is phenomenally superior to ours. Some say 10,000 times better, some say more like 100,000 times. It is an astounding difference and translates into an ability to sniff out one part per trillion of a substance. It means that sniffer dogs can pick up the trail of a person that passed through an area two or three days previously, an astonishing feat.

We cottoned on to the fact that dogs' sense of smell vastly exceeded our own pretty early on in our shared history and have made good use of it ever since. When ancestral humans formed an alliance with an ancient wolf species, some 15,000–40,000 years ago, we formed an impressive hunting partnership, combining the wolves' ability to track and hunt prey to the point of exhaustion with our weaponry to finish them off. Since then, dogs have been used to help humans throughout the ages. The police sniffer dog first came into being in 1895, when the French police force trialled an experiment with dogs accompanying foot patrolmen. It was an immediate success on the tough, gang-run streets of Paris and the German police quickly followed suit, using Alsatians and Doberman Pinschers to great effect. During the First World War tens of thousands of dogs were placed into the field, including 30,000 by the German army alone.* By 1939, it was estimated that the Nazis had approximately 50,000 Doberman Pinschers, sheepdogs, Alsatians and rottweilers trained for active service, including as pack carriers, first aid scouts and messengers. The British army enlisted dogs to sniff out landmines, including Rex, a fearless black Labrador who is said to have sniffed a clear path through the Reichswald forest for Winston Churchill.

Today, dogs are still used to detect landmines in war-torn countries, such as Sudan, Afghanistan and Yemen. They also help us to identify illegal substances, wanted persons and survivors buried under metres of rubble. En route to the US in 2018, I was delighted to pass through a special dog security check point. Other passengers and I on the Denver-bound flight had to proceed one-by-one past a highly alert and cheerful spaniel, who diligently sniffed each of our bags before we could enter the gate. More recently, dogs have been investigated for medical detection purposes, with evidence that they could identify patients with early-stage bladder, colorectal and lung cancer. It is thought that developing tumours produce certain compounds that dogs can detect in patients' breath or urine. Since the start of the Covid-19 pandemic, some centres have found evidence that dogs might be able to identify people who are infected with the virus. Dogs are even used in conservation: specially trained sniffer dogs are used to track down individual kākāpō when their

---

*Indeed, the Briard (an ancient breed of French herding dog) was used by the French army for Red Cross, sentry and ammunition-carrying duties and came close to extinction following the losses of the First World War.

transmitters fail; a job of vital importance when only 204[*] of these large, flightless parrots are left in the world.

Dogs are capable of perceiving things that we couldn't even imagine – and that's the problem. In our vision-based world, what we see and what we do are so intimately tied together that when we started investigating cognition in other animals, it was only natural that we approached it from a visual perspective. One analysis of the 253 studies on canine cognition published between 1965 and the end of 2012 found that only nine per cent focused on olfactory manipulations, while 74 per cent were visual. And while dogs are far from blind, there are some significant differences between their vision and ours.

* * *

Vision is a fantastically complex sense, involving the translation of millions of pieces of information every second from light waves into a meaningful image. These light waves are from the visible part of the spectrum, the 'Richard of York Gave Battle in Vain' / 'Roy G. Biv' part, rather than the parts that sit either side (ultraviolet or infrared) or even further away (gamma or X-rays). Visible light waves enter the eye through the pupil and are refracted by the curved cornea so that they are focused on to the retina, the sensory tissue at the back of the eye that begins coding the information. Like a pinhole camera, the picture that hits the retina is upside down. The retina is packed with millions of receptor cells called rods and cones, and its job is to translate the picture into messages about colour, shape and motion, sending these to the brain via the optic nerve. Rods and cones contain special pigments that convert the information in different

[*] As of May 2021.

ways. In rods, the protein only functions in low-light conditions. Cones, on the other hand, contain photopigments that are sensitive to different wavelengths of light, allowing the animal to see all combinations of colour possible with the cone types it possesses. Cones also enable the perception of fine detail. They don't work at night, meaning that once darkness falls our world loses both colour and clarity.

We have three types of cone ('trichromatic') and they correspond to the red, blue and green parts of the colour spectrum. Dogs, on the other hand, are dichromatic; their two cone types contain photopigments that correspond to the green-yellow and blue parts of the spectrum. This doesn't mean that dogs only see green-yellow and blue, just as we don't only see red, blue and green. But it does mean that the dog's visual world is more limited than ours, and they're much more sensitive to colours in these parts of the spectrum.[*] In fact, this is the case for the majority of mammals, and reflects our evolutionary history. Colour vision is thought to have evolved some 200 million years ago, in prehistoric vertebrates that possessed four different types of cone. This was long before mammals appeared and once they did, ancestral mammals lost two receptor types, probably because they were nocturnal and relied more on smell than vision. That makes us, with our three cones, as well as apes, Old World monkeys and at least a couple of other primate species, quite unusual mammals. The current consensus is that the third cone may have re-evolved as this lineage became more active in the daytime, possibly as an adaptation that helped with distinguishing ripe red fruit against the green of the forest.

---

[*] This is very similar to a human with deuteranopia, or red-green colour blindness, who lacks cones with pigment that matches red or green light.

Back to dogs: in addition to reduced colour vision, they are also about twice as bad as us at distinguishing changes in brightness and, depending on the study, anywhere from three to eight times worse at perceiving detail. Dogs are generally considered as near-sighted, although only by our measure of testing. A dog would never be capable of reading the top line on a sight test, but not because they lack the ability to focus at distance (they'd never excel as hunters if that was the case), rather, they are poor at differentiating closely spaced objects ('visual acuity'). One popular estimate puts dogs' acuity at about 20/75, meaning that from 20 feet (6m) away, a dog could perceive an object that a person with normal vision could differentiate from 75 feet (23m) away. In tests with people, images that were adapted to approximate dog vision (i.e. with fewer colours and more blurring) were harder to process than those that were normal.

Dogs have set aside a smaller proportion of their brains to process visual information and fewer nerve cells transmitting inputs from the eyes to the brain. In fact, not only is the region of the brain that processes inputs from the eyes considerably smaller than the region that processes inputs from the nose, it's also smaller than that for processing sounds. Dogs also have far fewer cones than humans, which translates into superior daytime vision for us. And yet, dogs' visual systems are well-adapted to their lifestyles. They're much more sensitive to movement at a distance than we are and the central region of their retinas is densely packed with rods as well as cones (we only have cones), which gives them superior twilight vision. Low-light vision is also enhanced by a specialised reflective layer in the eye called a tapetum. This is what makes your dog's eyes glow eerily if you shine a light at them in the dark: the function is to give the retinal cells a second chance of being stimulated by re-reflecting light.

Dogs' lack of reliance on visual information is precisely what we should expect for a mammal. When we talk about dogs 'failing' the mirror test we should remember that by downgrading our ancestral superiority in the sense of smell, we, along with the rest of the primates and the cetaceans, became the odd ones in the mammalian world. Dogs remained true to their evolutionary past by following their nose over their eyes, meaning that the mirror task may never be an appropriate experimental test for them, and many other mammals too. In fact, mirrors are probably downright confusing for dogs when they first encounter them. They see another dog and they can get close to it, but not close enough to do what they would do in nature, which is have a good sniff of its backside to get valuable information. In fact, they don't get any scent cues at all. Visual information can be useful: body stance, for example, provides important information about how the other animal might behave. But here, too, the intruding animal is odd. If Fido wants to communicate his dominance to the strange, scentless animal, he might growl, stiffening his body and raising his tail, only to be met by exactly the same. In nature, this would be unusual: aggressive displays are usually quite ritualised, and an opponent should respond by communicating its submission or its intention to escalate to a fight. Identical behaviour isn't helpful for an animal that relies on knowing its place within the group. The dog therefore learns that no matter how long it growls and stiffens its body, nothing is going to happen; changing tack, it also learns that no matter how much it playfully prances in front of the other creature, they aren't going to join in. Dogs may well habituate to the mirror being another meaningless feature of their environment.

* * *

In terms of cognitive ability, dogs have had a bit of a rocky ride over the past century or so. In the late nineteenth century, they occupied an elevated position among animals for their perceived emotional and cognitive abilities and anecdotal reports of their intellectual prowess abounded; for example, in 1884 physiologist George Romanes wrote: 'The emotional life of the dog is highly developed – more highly, indeed, than that of any other animal. His gregarious instincts, united with his high intelligence and constant companionship with man, give to this animal a psychological basis for the construction of emotional character.'

As we have heard already, the anecdotal approach fuelled a counter-movement and the emergence of behaviourism, with a focus on quantifying how environmental stimuli shaped the production of behaviour. Because of dogs' close evolutionary relationship with humans and the fact that they exhibit such a wide range of social behaviours, they were taken into the lab as subjects of early behaviourist studies, with a specific focus on their relevance to understanding human behaviour. As Ádám Miklósi, professor of ethology at the Eötvös Loránd University in Hungary, has written, this understanding of the dependence of dogs on humans makes the lack of concern for their welfare in the resulting experiments quite astonishing. 'Similarity in cognitive states was not extrapolated to similarity in emotional states. Quite simply, the dogs' suffering was not of much concern.' Some of the experiments referred to by Miklósi involved administering electric shocks and then recording whether the animals would learn the correct behaviour to avoid the shocks. Rats learn this easily, but dogs, surprisingly, struggled. After the first few shocks they seemed to give up and accept their fate, and their passive helplessness was equated with human depression. Miklósi points out that there is no natural analogue for dogs to receive such painful shocks; the closest situation may be when they receive a bite from another dog during an aggressive encounter. The dog's natural behaviour in these cases would be to try and avoid further attacks by adopting a submissive posture, which may have been why they behaved in a way that to us looked like 'giving up'.

Over the past 30 years there has been a huge increase in the amount of research conducted on dog cognition and today many universities have dedicated canine cognition labs. Miklósi established the Family Dog Project in 1994, the first such group to focus on the cognitive aspects of the dog–human relationship. There was no dog research going on before this and Miklósi has commented that his first journal article submission was met with: 'Who cares about these animals – why is it interesting to write about this stuff?' Given how prolific the field of canine cognition is today, it's staggering to think that dogs were considered uninteresting for cognitive research just a couple of decades ago, although it mirrors the experience of the earliest corvid cognition researchers too. For dogs, it came down to the prevailing view that domestication had corrupted their behaviour, breeding out any interesting natural abilities. Prominent figures dismissed domesticated species as cognitively inferior shadows of their wild ancestors, meaning that these animals were considered neither interesting nor relevant for studying intelligence. In fact, canine cognition is one of the hottest fields in animal behaviour research today.

As Range pointed out in the previous chapter, dogs have a different form of intelligence to wolves – one in which human influence plays a prominent role. We have already seen how dogs used a different strategy to wolves when it came to tackling a difficult problem, and cleverly designed experiments from dog cognition labs have provided a wealth of evidence of just how well dogs have adapted to living with people. For example, Miklósi and colleagues were the first to investigate whether dogs make use of human cues: a topic which at that point was talked about only to highlight poor experimental design. It harks back to the case of 'Clever Hans', the wonder horse that was said to be able to count (more in the next chapter), and the salutary lesson that experiments must be designed well enough to eliminate potential human cues that might influence the animal's behaviour. Miklósi thought about it differently: for dogs, at least, sensitivity to human cues is precisely what we should expect, so rather than trying to eliminate them from an experiment, much could be learned by studying how dogs use them. People's behaviour was therefore crucial and the team set about asking whether dogs could use a scale of different cues to locate a treat that was hidden under one of two identical pots. They found that dogs could utilise pointing, bowing, nodding, head turning and glancing gestures provided by their owners to find the food. It was the first study to show, convincingly, that another species could use human communicative cues.

Since then, dozens of experiments have been conducted on this topic and the emerging picture is that dogs are highly tuned to even subtle changes in our behaviour. As well as making use of our gaze direction to find food, dogs use their gaze to communicate with us. We have already seen that they look to us when they need help on a problem, and they also alternate their gaze direction when they want to convey something (for example, looking between us and the door if they want to go out). Some experiments suggest that dogs may also be sensitive to what we know, as well as what we do, which relates back to the tricky question of theory of mind. In one experiment, dogs were separated from an experimenter by a screen, one half of which was transparent and the other opaque. On the dog's side of the screen, one toy was positioned behind the transparent side (visible to the experimenter) and the other behind the opaque side (invisible to the experimenter). When the experimenter gave the command, 'Bring it here!' the dog more often selected the toy that was on the transparent side of the screen, indicating that it knew that was the only toy the experimenter could see. When the experimenter's back was turned, or she was sat on the same side of the screen as the dog, the dog showed no preference over the toy that it selected.

In another experiment, a dog was trained to expect a food reward under one of several buckets, with a human 'informant' indicating which one by pointing. Once the dog was reliably using the informant's signal to choose the right bucket, the task changed and two different informants signalled towards the buckets. One of them was facing the buckets when the food was hidden and therefore had knowledge of the correct bucket, while the other had their back turned to the buckets so had no clue where the food was hidden. If dogs have a 'theory of mind' and could think about what each of the informants knew, they should choose the knowledgeable person. They did. But that's not the only explanation – dogs could make the same choices by simply learning that the person facing the food when it was hidden was a better predictor of food than the one that had faced the other way. As we saw in the previous chapter, theory of mind studies are contentious, because there is a crucial difference between understanding that others experience the mental state of 'seeing' and, on the other hand, learning that the direction someone looks in is important. And, unfortunately, it's really difficult to tell them apart. As yet, the evidence is inconclusive in dogs, but it's something that multiple canine research groups are working on.

When it comes to mirrors, other research indicates that dogs can at least learn to use them. Back in the late 1960s, Michael Fox, Bekoff's mentor, used mirrors to study dog development. He took young Chihuahua puppies and allocated each one to either be raised in a litter with other puppies, or in a litter with cats. The cat mothers seemed unfazed about fostering a little puppy alongside their kittens. Until the age of 16 weeks, each Chihuahua had only the company of other puppies or kittens. To compare the behaviour of the two groups, Fox then presented each with a mirror. The differences were remarkable. All normally reared puppies were immediately interested and excited in the mirror: they barked at it, wagged their tails, jumped up and pawed at and around it. The cat-raised pups acted completely differently: they either ignored the mirror or treated it with utmost caution, approaching silently and gingerly, with their tails tucked between their legs. These puppies had never seen another dog before; they had no mental image of what a dog was, how strange it must have been! When the two litter types met, the dog-raised pups played normally while the cat-raised pups huddled with their kitten-litter mates. They didn't seem to know how to be a dog. Yet, two weeks after all the pups were housed together, the cat-raised Chihuahuas were behaving normally: playing with each other and responding to a mirror as if it was a playmate.

Some researchers have explored dog interactions with mirrors further, asking not whether they recognise themselves, but whether they can use them as problem-solving tools. The results are mixed. Dogs required habituation to the mirror, just as Gallup's chimps had done, which makes sense: when the function of a tool is not visibly obvious (like a hooked stick), animals must learn its properties before we can expect them to apply it to a problem. Nonetheless, some dogs did successfully use the mirror to locate their owner, who was otherwise out of sight, and some dogs similarly learned to use it to locate food. Mirrors clearly do not play an important role in a dog's life, but they're perhaps not entirely meaningless either.

★ ★ ★

Let's return to Jethro, snootling* around in the vegetation of Boulder Creek. Bekoff and Fox were intrigued by the negative

---

*Snootling is a word I grew up with, which came to be known as a 'Wimpenny word'. For me it perfectly sums up the behaviour of a dog (it's generally always used with dogs) sniffing round in the undergrowth, nose buried into vegetation and earth, inhaling the smells of nature.

mirror-recognition results, so they started thinking about an alternative way to test for canine self-awareness and thoughts turned to smell. It was clear from their observations at dog parks that individuals spent far more time sniffing other dogs' urine than their own. Bekoff initially wanted to observe how wild coyotes behaved when they encountered the scent of other animals, but there were too many practical difficulties. Watching Jethro and other dogs along Boulder Creek one day, he had an inspiration. Why not just move the yellow snow and see how the dog responded? For the next five years, whenever there was snow on the ground, Bekoff approached Jethro's early morning walk as a data-collection opportunity. Bekoff concluded that 'Jethro had some olfactory sense of "self", a sense of "mine-ness" but not necessarily of "I-ness"'.

When Bekoff published his findings with Jethro in 2001, he set in motion a new approach to studying self-awareness in dogs. The idea inspired Horowitz to investigate further. Enrolled on a cognitive science PhD programme, Horowitz became interested in what natural play behaviour could reveal about animal minds. She hadn't planned to study dogs – the field didn't exist at this time – but she owned a dog, and was inspired by a single line in a book by Bekoff and the philosopher Colin Allen about how play behaviours might give insight into dogs' understanding of intentions. 'Dog play is terrific, and they turned out to be good subjects to study,' says Horowitz. 'After I finished my dissertation, I got more interested in the animal than the theoretical question that brought me to them'. Since there was no one at her institution working in dog cognition, Horowitz reached out to Bekoff for advice on how to observe dog behaviour and he became her mentor. For Horowitz, Jethro's behaviour clearly indicated that the animal recognised a difference between its own odour and that of other animals. But questions (and critics) remained, so Horowitz set out to learn more. As a student she had learned about Gallup's MSR test and the exciting possibilities it offered for studying a difficult question. She decided to develop an analogous test for dogs, but with an olfactory, as opposed to visual, mirror. Her aim was to experimentally evaluate whether dogs recognised the smell of their own urine as themselves, in the same way that chimps recognised their own mirror reflection. Instead of the animals having a mark applied so that their visual reflection changed, the key test here was to see how the dog reacted when there was a change to its olfactory reflection. Would dogs spend longer sniffing at samples of their urine that had been modified than at unmodified samples?

The study took place at the Port Chester Obedience Training Club in New York. Thirty-six dogs visited the centre, including at least 18 different breeds: from the tiny hairless Chinese Crested to large Belgian Tervurens and German Shepherds. Owners were provided with a sample cup and a vinyl glove ahead of the test, and instructed to collect a fresh urine sample from their dog before they arrived. On each trial, the dog was let into a testing arena in which were placed two canisters. The difference between the canisters was what was inside, which fell into one of four groups: 'self' was just a small quantity of that dog's own urine; 'self-modified' was that dog's urine containing a small piece of diseased spleen;[*] 'other' contained the same quantity of urine from an unknown dog; and 'modified-only' just contained water with the spleen sample.

The dogs spent considerably longer sniffing at both the 'modified-self' canister and the 'other' canister than at the canister containing their unmodified urine. This seems consistent with Bekoff's results as it suggests dogs were less

---

[*] The spleen sample came from a dog with lymphosarcoma; since dogs have shown sensitivity to diseased tissue, this was intentionally chosen for its potential salience.

interested in their urine compared with something that was different. However, the dogs were also very interested in the canisters that just contained spleen, suggesting that they may simply pay more attention to odours that are novel or interesting. The team ran another set of trials, this time with the more neutral modifier of anise oil.

This time, the findings were more clear cut. The 11 new dogs tested spent significantly longer sniffing at their modified urine compared with both their unmodified sample and the anise oil alone. There seemed to be something compelling to the dogs about a smell that was both their own and different. Was it that they recognised their own scent had been changed? Horowitz and colleagues were cautious in their conclusions, proposing that the study could be considered analogous to previous mirror tests and suggesting that it provided evidence for 'partial' self-recognition. Like Bekoff, Horowitz emphasised the necessity for comparative cognition studies to be designed around the perceptual abilities of the species being tested.

Horowitz had expanded the study under captive conditions, imposing experimental controls that were impossible for Bekoff to implement out on the trail. 'I couldn't do what they did with free-ranging dogs,' reflects Bekoff. 'But the nice thing was, we all agreed!' Gallup, on the other hand, while considering Horowitz's study a 'compelling analogue to the mark test', remains unconvinced that the experiment provides evidence of self-awareness. He points out that a crucial part of the mirror test is missing from the study: chimpanzees used the mirror to explore the mark on their bodies, but there was no evidence that the dogs went from sniffing their modified scent to sniffing themselves. 'That's an interesting idea,' agrees Horowitz, 'but since the mark wasn't on the dogs, I don't know why (or where) they would smell.'

Where do we go from here? It's clearly a controversial topic and something more seems to be needed to resolve the opposing sides. For Bekoff, the gel that will glue these behavioural findings together is neuroimaging. It offers a way into the mystery of what it is to be a dog and it's difficult to argue against images of brains that show activity corresponding with behaviours.

Gregory Berns is a neuroscientist and professor of psychology at Emory University in Atlanta, Georgia. His stance is that behaviour is important, but it doesn't tell the whole story, so he's on a quest to show that we can understand how animals experience the world by scanning their brains. Berns and his team train dogs to enter an fMRI scanner and lie down motionless: remarkably, the animals are awake, unsedated, unrestrained and free to leave at any time. One study asked how they responded to smells. Twelve dogs were presented with five different smells: self, familiar human, unfamiliar human, familiar dog and unfamiliar dog. The team focused on analysing responses in a region of the brain called the caudate nucleus, which is part of the reward system and rich in dopamine receptors;* in humans, this region lights up in anticipation of a treat. The scans showed caudate activation in dogs too, with the strongest response to the smell of a familiar human. It's tempting to think this reveals an emotional attachment, but another possibility is that they simply associate their owner with food. A follow-up study was devised to try and tease apart the two options: are dogs in it for food or love? This time, 15 dogs were trained to associate different toys with either a food treat or praise. The scanner then recorded activation levels in the same brain region when each toy was presented alone. Most of the dogs had equal responses to both, with some responding more

---

*Dopamine is a neurotransmitter (i.e. a chemical messenger that transmits information between neurons) and is popularly known as the main 'feel-good' chemical. It is released by the brain during pleasurable activity and is important for arousal, memory formation, learning and pleasure. It also has lots of other effects within the brain and around the body more generally.

strongly to the toy associated with praise than food. It provides powerful evidence to support many owners' beliefs that their dogs are genuinely attached to them.

With these advances in neuroimaging and our increased knowledge of dog cognition, one tantalising thought is whether one day we might be able to see evidence for self-awareness in the brain. I put this to Berns and he is quick to bring me back to Earth: 'It is possible we could use neuroimaging to answer the question, but it would require a solid basis in humans – which is the only species that we know it exists in and that we can image ... We don't know what that would look like in dogs or whether it exists at all.'

Fact or fiction?

Aesop's dog seems to be grounded in fact – there is no evidence that dogs or any other canid can recognise their reflected image. Since Gallup pioneered the MSR test more than 50 years ago, it has stimulated a huge amount of research and there is convincing evidence that a small number of species do recognise themselves in the mirror. The thornier question is what this reveals about self-awareness; that is, are these animals truly becoming the object of their own attention, as we do when looking at our reflections? And, conversely, what does failing the mirror task reveal? It certainly seems plausible that animals, like dogs, which rely more on smell for communication and recognition of other animals, would not behave in the same way to their reflected image as animals that predominantly use vision.

Another question concerns the supposed relationship between self-recognition and theory of mind. As we heard in the previous chapter, it's unclear whether the evidence for theory of mind in apes and ravens will ever match the full-blown 'mindreading' of humans. However, given their different evolutionary histories, one might wonder why we would expect that it should. And if this is the case then perhaps the relationship between self-recognition, self-awareness and theory of mind is not as clear cut as suggested. Here's Thomas Bugnyar again: 'For us, it's logical that these things go together, but from a biologist's point of view, I'm not so sure. I think the less we come with preconceptions, the better. The more we test empirically, the better. And the more these tests are not depending on one paradigm, the better!'

Gallup's mark test has been the single, gold-standard method for evaluating self-awareness since 1970, a fact that demonstrates the buy-in it has had from the scientific community, who largely agree with the rationale and are convinced by the chimpanzee and orangutan behaviour. But hinging everything on one test will always be problematic: alternative paradigms are needed to explore the questions in other species and enable the construction of a bigger picture. Whether alternative types of experiment can tap into the same cognitive abilities is a matter for debate, but in recent years some alternatives have been put forward, which have the potential to reveal different aspects of an organism's sense of self. The olfactory mirror is one such avenue of research. Another is body awareness. Berns explains that all animals must be able to distinguish themselves from the rest of the world: this is how they can move through their environments efficiently, without constantly bumping into things. 'From the simple differentiation of me/not me, we can begin to think of different levels of self-awareness,' he says.

One way to study this simpler form of self-awareness is to ask whether animals recognise their own body to be an obstacle. Using an approach first developed with infants, this was investigated in a study of Asian elephants. The elephants were required to walk on to a rubber mat, pick up a stick and pass it forward to the experimenter. In control trials, the stick was freely available on the mat, while in test trials it was attached by a short rope to the mat, meaning that the mat had to be passed with the stick. The question was, would the elephants recognise that they needed to step off the mat to succeed in the test trials? The results suggested that they might, leading the study authors to conclude that elephants 'may have some specific awareness of the relationship between the self and external objects'. More recently, an equivalent study has found something similar with dogs, who were required to pass a toy that was either attached to the mat they were standing on, or the floor. It's a deceptively simple test, but potentially fits a piece to the puzzle of what kinds of data may reveal self-awareness.

It's clear this topic will continue to prompt strong opinion for years to come. Like Bekoff, Horowitz thinks that multiple approaches are needed, spanning behaviour and brain, and she explains that we're only just starting to understand dog self-recognition. 'There are so many unanswered questions! I am still very keen to expand what we know about dogs' olfactory experience: about how they perceive themselves and others through smell. My lab is working on those questions now.' Undoubtedly, there is a lot more to learn before any firmer conclusion can be made. Aesop's dog may not understand his reflection, but that doesn't necessarily mean he doesn't have a sense of self.

CHAPTER FOUR

The Ass Carrying the Image

An Ass once carried through the streets of a city a famous wooden Image, to be placed in one of its Temples. As he passed along, the crowd made lowly prostration before the Image. The Ass, thinking that they bowed their heads in token of respect for himself, bristled up with pride, gave himself airs, and refused to move another step. The driver, seeing him thus stop, laid his whip lustily about his shoulders and said, 'O you perverse dull-head! it is not yet come to this, that men pay worship to an Ass.'

Three hundred miles south of Cairo, in the low desert west of the river Nile, sits the ancient sacred city of Abydos. Famous as a burial place for the earliest Egyptian kings and an epicentre of worship for the god Osiris, Abydos is one of the most important archaeological sites of ancient Egypt. In 2002, it yielded a discovery that nobody was expecting. Archaeologists were excavating 5,000-year-old brick tombs adjacent to the mortuary complex of one of the founder dynasty Egyptian kings. The tombs adjacent to other early kings at Abydos typically contained the remains of high-ranking courtiers – it is thought that their physical proximity signified their status and relationship to the king. Other adjacent tombs contained the remains of lions and several large, wooden boats were buried in another.

At this king's funerary complex, the excavated tombs did not contain the skeletons of courtiers, lions or other, typically symbolic items. They contained donkeys.* Ten donkeys, to be exact. They had been carefully placed into the tombs on reed mats, all oriented on their left sides and facing south-east. The skeletons were almost complete and wonderfully preserved, in places there were even remnants of hair and soft tissue. It was an astonishing discovery. Analysis of the bones showed signs of wear: these had been working animals that had carried heavy loads. Nonetheless, they were in good shape; they'd been well looked after. In fact, during the ancient Egyptian pharaonic times, donkeys enjoyed high status and had a distinguished role in society. It's possible that the Abydos donkeys were used to provision the royal

---

*Let's start by getting straight on terminology, because Aesop's fable refers to an 'ass', whereas I've already started talking about donkeys. They're the same thing. Ass comes from the Latin subgenus name for donkey, which is *Asinus*. This is now primarily only used in academic settings, because as the English language changed over time, 'ass' took on the meaning of 'arse', giving rise to a variety of commonplace sayings and insults. From about the 1930s in the US this became common language, but some attribute the first use of it to Shakespeare, if his transformation of Nick Bottom into a donkey in *A Midsummer Night's Dream* (1594) is the wordplay some believe. Here, I stick with donkey.

household and were considered to be so valuable that they were deemed worthy of accompanying the king into the afterlife.

Fast forward to 2010 where, in the Black Sea resort of Golubitskaya, a donkey was forced to parasail above the tourist-filled beaches of southern Russia. This donkey was called Anapka and she had been loaned to a couple of 'entrepreneurs' who said they wanted to offer donkey rides. Instead, they used her to advertise their water sports business. Anapka was strapped into a harness and unceremoniously hoisted into the air, dangling from a parachute that was attached to a speedboat. As she flew, nearly 50m high, Anapka brayed. The holidaymakers below covered their mouths in horror. Children cried. Overall, it was universally condemned as a terrible idea, but nobody reported the act and nobody was prosecuted. Anapka died the following year of a heart attack; the vet who examined her said she had probably never recovered from her ordeal.

Poor Anapka's tale is just one in a long list of crimes against donkeys. Over the past millennia these animals, once valued so highly that they were buried alongside royalty, became creatures of such lowly rank that people stopped believing they deserved respect or concern for their welfare. Donkey became the proverbial 'dumb ass', a symbol for ridicule. Many of Aesop's fables include donkeys and the animal's fate is usually to be eaten or beaten; an outcome that we are led to believe it deserves for being so easily tricked or for having the temerity to aspire above its lowly situation. When a fable includes a donkey, you know that it will not end well for the animal; the donkey is the fall guy, the very symbol of foolishness.

How we represent animals in fictional stories reveals our attitudes towards them and our treatment of them. It begs the question of what happened to change our estimation of donkeys from valued to vilified, respected to ridiculed. And are we correct to be so derogatory about donkey minds? What does science have to say?

` ★ ★ ★

It's hard to convey just how important domestication has been in the history of planet Earth. It resulted in a remarkable transformation in the lifestyles of early human societies that changed the course of human evolution and has enabled spectacular population growth. It is no exaggeration to say that we rely on domesticated species for our survival.

Most researchers agree that domestication was not a straightforward, linear event: for every domesticated plant or animal group, there were likely multiple, independent domestication events from different populations around the world and at different times. With that caveat in mind, a rough timeline of domestication would start with the dog which, as we've already seen, was thought to have domesticated itself from an ancestral wolf species at least 15,000 years ago. It is thought that farming crops first began some 12,000 years ago, in the Fertile Crescent of the Middle East, and a little later people also began keeping animals. First were sheep, descended from the Asiatic mouflon in the region of 10,500 years ago and shortly after came goats. Cattle and pigs were next, descending from the now extinct wild ox, the aurochs, and the Eurasian wild boar some 9,000 and 8,000 years ago, respectively. There is some debate about the timing of donkey domestication, but the earliest archaeological evidence places it as far back as 6,500 years ago and this was closely followed by horse domestication in the region of 5,500 years ago.

Donkeys belong to the same biological family, the *Equidae*, as the horse, three zebra species, the Asiatic wild ass (aka onager or hemione) and the Asian ass known as the kiang. They are descended from the African wild ass, of which, until relatively recently, there were two living subspecies.

The Nubian wild ass has not been recorded in the wild since the 1970s, so is today considered extinct. Photographic records depict an animal that is practically indistinguishable from domesticated donkeys, even down to the characteristic cross on the shoulders and back. The Somali wild ass, which is critically endangered,* looks a little different, lacking the cross and with the addition of striking zebra stripes around its lower legs. It was therefore long assumed that modern donkeys were descended from the Nubian wild ass, but genetic analyses were needed for confirmation and they revealed a more complicated story. Modern donkeys can be grouped into one of two genetic lineages (haplogroups) depending on their mitochondrial DNA. These groups are referred to as Clade 1 and Clade 2 and indicate that two independent domestication events must have occurred from different populations. An important analysis of donkey and wild ass DNA published in 2011 confirmed that Clade 1 donkeys are descended from the Nubian wild ass. The ancestor of Clade 2, on the other hand, remains unknown. The results ruled out any role of the Somali wild ass, leaving the door open for another, now extinct, subspecies.

The earliest archaeological evidence for donkeys being used as beasts of burden comes from the Neolithic settlement of El-Omari in pre-dynastic Egypt, dated 4600–4400 BC. It is thought that nomadic farmers in the Egyptian Nile Valley formed an alliance with donkeys' ancestors for pack transport, with one hypothesis suggesting this was a response to the region's increasing aridity. Donkey domestication happened slowly: some 1,500 years later, while the skeletons from Abydos show that the animals were being used to carry heavy

---

*The IUCN estimated just 23–200 mature individuals in 2014, with a decreasing population trend. The animals are confined to the eastern edge of the Horn of Africa, with confirmed reports in Eritrea and Ethiopia, and possible presence in Djibouti, Egypt, Somalia and Sudan.

loads, in other respects there was little difference from skeletons of their wild ass ancestors.

Donkeys have historically been valued for many reasons and by people from all walks of life. They were bred for meat and milk, but they were also the first animals to be ridden, which conferred a hugely elevated status to the rider compared with those on foot. Before camels were domesticated (some 3,000 years ago) and became the 'ships of the desert', donkeys enabled families and communities to travel further and with more possessions than they could on foot, enabling them to disperse and settle in more suitable areas. Subsequently, they also played a hugely important role in the development of trade. Caravans of donkeys transported metals, textiles, spices and other commodities hundreds and even thousands of miles. Historian Peter Mitchell has noted that many iconic historical objects were made possible by donkey labour. Tutankhamun's funeral mask, for example, was made from gold transported by donkeys from Nubia and turquoise from the Sinai; the lapis lazuli from Afghanistan (over 5,000km away) and obsidian from highland Ethiopia were likely transported at least partway by donkeys too.

From Africa, the donkey spread east, into the Levant, throughout the ancient Near East and onwards. Archaeological records show they were transporting olive oil and other goods in Palestine, around 3000 BC, and clay cuneiform tablets document the use of donkey caravans by Assyrian merchants for the transport of large quantities of tin and textiles between Assur in Mesopotamia (modern Iraq and Syria), and the ancient Anatolian city of Kanesh. As well as providing the means for merchants to trade their goods, donkeys were an important trade item themselves and were often sold upon the caravan's arrival.

Donkeys continued to spread across the globe and Mitchell has noted that by 1900 BC, not having a donkey was a sign of dire poverty. The Romans are thought to have relied on the animals to transport large quantities of goods as they extended their empire into Europe, Africa and Asia. That means that where the Romans went, donkeys went too: the animals arrived in Britain when the Romans invaded in 43 AD.

When archaeologists opened the tomb of Cui Shi, a noblewoman who died in 878 AD in Xi'an in central China, they found that it had been looted of everything but some animal skeletons, a lead stirrup and some other objects. These proved to be extremely interesting. At least three of the animals were donkeys and analysis of the chemical signatures left in the bones ('isotopes') revealed they had been well fed. Analysis of the bones and joints showed that these donkeys were fast runners and able to turn quickly, rather than animals burdened with heavy loads. They were Cui Shi's polo animals and this wasn't a one-off – several Chinese artworks depict high-status women playing polo on donkeys. Polo became incredibly popular in China during the Tang Dynasty, and it was a high-stakes, high-risk activity; researchers think that women may have chosen to ride donkeys because they were sturdier and safer than horses.*

Other attributes of donkeys were valued too. The Greek physician Hippocrates, known as the 'father of medicine', wrote about the virtues of donkey milk in the fourth century BC, prescribing it for all kinds of complaints: from fever to joint pain, liver problems to snake bites and poisoning. Donkey milk was also valued for health and cosmetic benefits. Cleopatra must have been convinced, since she kept a stable of hundreds of jennies to supply her with milk for bathing. In his encyclopaedia of the natural world, the *Historia Naturalis,* Pliny the Elder summarised the perceived

---

*Indeed, as a young girl, Queen Victoria learned to ride on a donkey called Dicky, who was decorated with blue ribbons, and in later life she preferred to drive a carriage drawn by a donkey, rather than a horse.

softening and wrinkle-effacing properties of donkey milk, writing that 'some women are in the habit of washing their face with it seven hundred times daily, strictly observing that number'.

The use of donkey milk has continued to modern times. In seventeenth-century France, babies whose mothers could not nurse them were provisioned with donkey milk, sometimes straight from the teat. In fact, analyses show that donkey milk is nutritionally similar to human breast milk, save for a lower fat content. Supplemented with a little extra oil, however, it makes a good alternative for children who are allergic to cow's milk. It is also high in immunoglobulins, and for this reason is used in Sudan as a native medicine to treat whooping cough and in India for the mosquito-borne disease chikungunya, although no clinical studies have been conducted to demonstrate its effectiveness.

Donkeys have been our beasts of burden for thousands of years and they continue to provide crucial labour for many communities around the world. The Food and Agriculture Organization of the United Nations estimated in 2018 that there were 116 million working equids worldwide, comprising 46 million donkeys, 60 million horses and 10 million donkey-horse hybrids. An estimated 85 per cent of the world's equids are found in low or middle-income countries and, according to Brooke, the international charity that protects and improves the lives of working equids, around 600 million people worldwide depend on the labour and income generated from these animals.

They are long-lived animals, hence the expression 'donkeys' years', and during this time offer much value as dependable workers that will reliably get the routine, arduous tasks (i.e. the 'donkey work') done. A frequent misconception about donkeys is that they are 'not productive'. And certainly, if your only measure of productivity is milk, meat or wool production, the donkey won't be winning any titles – but neither will a tractor and I challenge you to tell any farmer that their tractor doesn't contribute to productivity. Instead, donkeys pull ploughs, carry crops, provisions and merchandise, haul rock and timber, and turn heavy millstones to produce flour. An average animal weighing 160kg can carry up to 50kg on its back and pull up to twice its bodyweight on level ground. Their ease of keeping, hardworking nature and sure-footedness makes them extremely valuable animals, particularly among communities living in arid lands, but also in mountainous regions of more temperate zones, where it would be impossible to use mechanised transport. In the Italian alps, for example, 'donkey nannies' are used to transport new lambs down from the mountaintops to fertile grazing land below; a trip that would be far too perilous for the young animals on foot. The donkey is fitted with a special blanket sporting lamb-sized 'pockets', offering a devastatingly simple, yet very practical, solution.

Donkey domestication transformed early pastoral societies. They facilitated previously unimaginable changes and socio-economic advances to those groups. Our timelines have been intertwined ever since, so much so that it is impossible to tease apart the donkeys' contributions to human history. So, the next question must be – where did it go wrong?

About the same time that the donkey was becoming domesticated in north Africa, its larger, faster cousin was forming a relationship with people thousands of miles away. Horse domestication is a complex topic because, as with the donkey, it is probable that more than one domestication event occurred, making it difficult to pin down its ancestors. The timing is also a challenge, because there will have been frequent interbreeding by horses in the early stages of domestication and their wild counterparts. The European wild horse (Tarpan)[*] is considered the most likely ancestor, although other subspecies of wild horse may have been involved.[†] The current consensus is that the Botai hunter-gatherers in Kazakhstan domesticated wild horses some 5,500 years ago, but there were likely other events.

From around 2000 BC, domesticated horses were transported from their ancestral homelands into eastern Europe and south-west Asia, where their size, strength and power contributed to the developing civilisations of the ancient world. Donkeys had enabled societies to advance, but horses took them further – they could pull heavier loads, transport people more quickly and add more strength to an army. Wherever they went, horses took over as the premier animal for transportation, hunting, warfare and royalty.

There was an additional consequence to the horse's arrival. Within the equid family, all species also belong to the same genus, *Equus*. This lineage is thought to have arisen in the region of 4.5–4 million years ago, and not started to diverge until 3–2.5 million years ago. Consequently, all *Equus* species living today are closely related enough that they can successfully interbreed. Equid hybridisation had already

---

[*] As recently as 200 years ago, tarpans were found across Eurasia, and it is the tarpan that we see depicted in the stunning prehistoric cave art of Chauvet and Altamira.

[†] Przewalski's horse, the small, stocky breed found in Mongolia, is often considered to be the last living representative of the wild horse. After going extinct from Mongolia in the 1960s, a concerted conservation effort elevated numbers of Przewalski's horse enough to allow reintroduction in the 1990s, and the latest population estimate puts them at more than 2,000 individuals. However, a recent analysis of Przewalski's horse DNA shows that they are descended from the Botai domesticated horses. Sadly, this means that true wild horses went extinct hundreds, if not thousands of years ago. Przewalski's, mustangs and other apparently wild horses are in fact the descendants of feral animals that escaped domestication.

started in areas such as Anatolia in the third millennium BC: here, the hybrid offspring of a donkey and onager (called a kunga) was said to be stronger than a donkey and more manageable than an onager. These animals even saw the field of battle! 'The Standard of Ur', a Sumerian mosaicked box dated around 2500 BC, depicts elite Sumerian warriors going into battle in war-carts pulled by kungas, the bodies of their enemies trampled underfoot. Yet it was also around this time that people started breeding horses with donkeys, discovering that the offspring offered superior qualities to the kunga. Mules (if the father is a donkey, aka jackass, and the mother a horse, aka mare) and hinnies (if the father is a horse, aka stallion, and the mother a donkey, aka jenny) are infertile animals,* and they combine traits of both their parent species. The mule seems to have combined the most useful characteristics of both – it provides a classic example of hybrid vigour, where traits are enhanced as a result of the interaction between parental genes. This has led some to argue, quite unfairly, that the mule is the donkey's greatest contribution to human history. The mule's value is apparent from historical artifacts such as a price list for livestock dating to the Hittite period (between about the seventeenth and twelfth centuries BC), on which they were the most expensive species. The hinny is a smaller animal that is less celebrated; it is, however, also the least studied of all domesticated equids.

Horses and donkeys are well-represented in ancient writings and artifacts, but in quite different ways. Horses were expensive, high-status animals that symbolised wealth, power and beauty. They were integral to the economy, an essential part of warfare and featured heavily in mythology.

---

*This is because donkeys and horses have different numbers of chromosomes – donkeys have 62 pairs, whereas horses have 64 pairs. Mules and hinnies have 63 pairs, which means they cannot breed with either parent species and produce viable offspring.

Although still a domesticated animal, the way that horses have been depicted in Greek literature and art is worlds away from the donkey – something that historians have suggested may represent the difference between the upper and lower classes in Greek society. Horses are featured in multiple forms of art, including vases, sculptures, mosaics and even buildings. Donkeys do not adorn the sides of a precious vase, nor did noblemen commission sculptures of donkeys, and there certainly weren't any divine winged donkeys in Greek mythology. They survive in literature, but little of it is positive. Aesop alone wrote around 20 fables with donkey characters and they are invariably portrayed as lazy, greedy or stupid.

As times changed and people wanted different things from their equids, the donkey seems to have fallen out of favour. At least, that is, in the cities and civilisations from which we have surviving accounts. For nomadic people and in poorer, rural communities, donkeys remained essential, valued animals. There's a problem here, points out Faith Burden, executive director of equine operations at the Donkey Sanctuary in the UK,[*] in that surviving texts and artworks come from the places where horses had higher value, which provides a biased view of how donkeys and mules were perceived overall. She says: 'If peoples living in remote, arid,

---

[*] The Donkey Sanctuary, based in Devon, came into being in the summer of 1969, when Dr Elisabeth Svendsen, owner of a Devonshire country hotel, bought a jennie whom she named Naughty Face. Over 50 years later, the sanctuary has cared for more than 20,000 donkeys at its headquarters in Devon and at one of its many centres that have been established around the world. Svendsen's motto, that 'donkeys come first, second and third' has directly influenced the lives of the animals in the sanctuary's care and indirectly improved the situation for countless other donkeys by promoting education and understanding of these animals.

mountainous regions had been those able to record their thoughts in stories, etc. then perhaps the donkey and mule would have been more revered.' As it was, the money centres propagated their disregard for donkeys – from Greece to Rome and on, through time and beyond.

There was no improvement in donkey depiction by medieval times, as is apparent from the writings of thirteenth-century Franciscan monk Bartholomaeus Anglicus: 'For the elder the ass is, the fouler he waxeth from day to day, and hairy and rough, and is a melancholy beast, that is cold and dry, and is therefore kindly heavy and slow, and unlusty, dull and witless and forgetful.' Twentieth-century donkey characters include the gloomiest animal in the forest, Eeyore; the most cynical beast on *Animal Farm*, Benjamin; and the most annoying (albeit adorable) animal in Shrek's swamp, Donkey. Donkeys are still associated with stupidity, stubbornness and servility, although in many cases there's a big difference in perception between those that rely on the animals and those that only know them from stories.

Ill-treatment of donkeys comes down to two main drivers: poverty and a lack of knowledge. Poverty is the main one – the need to earn money from using the animals can result in terrible mistreatment, as Burden explains: 'If people can earn an extra few dollars through mistreatment of their animals, I'm afraid many may choose to do so as their existence is so finely balanced. In many of these cases the humans and donkeys live a similarly miserable existence and it is hard to separate the hardships from the behaviours.' Education is key, particularly in helping owners or workers to understand the longer-term impacts on productivity. This involves, first, addressing the commonly held misconception of the 'small horse with big ears', ensuring that donkeys are kept in conditions that suit donkey needs. Second, there is a need for better understanding of donkey behaviour: for their owners, to inform the ways in which they treat them; for veterinary professionals, to help interpret behavioural cues when they assess and treat them; and for the wider public, to improve popular perceptions of the animals.

Education programmes are helping and yet a shocking report from the Donkey Sanctuary in 2019 warned that the world's donkey population could be halved in just five years. That's a potential loss of tens of millions of animals. The threat comes from a booming international market for donkey hides, which are being sold to manufacturers of ejiao, a Chinese traditional medicine made from donkey gelatine. Ejiao used to be restricted to the elite classes, but has been rebranded for China's growing middle classes as something of a wonder drug. Demand has soared and is now outstripping supply, even with the massive price hikes that have accompanied its rising popularity. China used to have the world's largest donkey population, with 11 million animals in 1992. That's fallen to around 3 million animals today, meaning that suppliers have had to look to the rest of the world. Donkeys are being bought up and sent for slaughter in many of the countries where they contribute the most to livelihoods, particularly in Africa and South America. In Kenya, for example, the first donkey abattoir opened in 2016, backed by Chinese investors. Three more followed, with hundreds of thousands of donkeys bought up across the area – between 2016–18, 15 per cent of Kenya's total donkey population was slaughtered. It led to shortages and price hikes, and that's when the thieves moved in, stealing animals from the communities that rely on them most. It's a trade that has led to considerable suffering: for the animals, which are treated appallingly, and for the people whose livelihoods are stolen and who are then priced out of replacing their animals. The trade and the slaughter continues and looks set to unless tough restrictions are put in place or a synthetic, cheaper ingredient can be identified. The fact that these reports have not prompted a mass public outcry is perhaps revealing in itself.

* * *

Spend time in the company of a donkey and you'll notice a few character traits. There's a profound patience about these animals, a calmness in their large, dark eyes as they stand before you, seeking a treat or head scratch. Burden first visited the Donkey Sanctuary aged seven and was struck by how gentle the animals were. Years later, when she was offered a job there that combined her love of science with donkeys, it was a dream come true. Today, she is a globally recognised expert on the health and welfare of donkeys and mules, and heads up the sanctuary's considerable research and operational activity. And yet, 16 years later, she has not lost her affection for the animals: 'There is nothing quite like the gentle breath of a donkey on your face or the velvet muzzle of a donkey gently seeking a little human contact.'

One of many consequences of the belief that donkeys are 'poor relations' of horses is that less information exists on their welfare needs. Far more is known about the anatomy, physiology and behaviour of horses, particularly in places where horses are the dominant equid (which is much of Europe, the UK and US). With so little specific research into the needs and traits of donkeys and mules, most vets in these places have had to fall back on horse information. The trouble is, while horses and donkeys are closely related, 3 million years is plenty of time for some crucial physical and behavioural differences to have evolved.

To really understand these differences, we need to look to their respective evolutionary histories. The donkey is descended from African wild asses: tough, sturdy animals that evolved in the arid bushlands and mountainous regions of north-eastern Africa. The domesticated horse, on the other hand, is descended from the wild horses of the central Eurasian steppes. These vastly different habitats shaped the wild animals in distinct ways and these are preserved in their domesticated counterparts.

Let's start with the donkey's characteristically large head. Often a source of mockery, its head anatomy is actually all about food. Larger bones mean a more powerful jaw, an adaptation that enabled wild asses, which may have ranged over 20–30km every day, to expand their dietary options by chowing down on tougher, coarser plants and shrubs. Wild horses, on the other hand, adapted to the expansive plains of the steppes and a largely grass-based diet; there was no selection pressure for an extra-strong jaw. Donkeys are also superbly adapted to digest food of poor nutritional quality and their daily energy requirements are up to 50 per cent lower than a horse of similar size. In fact, donkeys that are provided with either too much food or food that is too calorific are at risk of potentially serious health issues.

Those big ears help to facilitate communication between scattered groups and dissipate heat in the hot, arid environment of northern Africa. These ancestral climate differences also manifest in the shelter-seeking preferences of donkeys and horses. One study at the Donkey Sanctuary found that donkeys preferred to be outside when it was hot and sunny, while horses tended to seek shelter when the temperature was above 20°C. Conversely, when the temperature dropped below 5°C or when it was wet and windy, donkeys sought shelter, while horses seemed to be OK outside. A related study conducted in southern Spain found that donkeys were less affected than mules by rising temperatures and light levels, although both sought shelter when the average temperature was about 26°C.

Donkeys also show several adaptations to aridity. For starters, they can go for two to three days with limited access to water and then rehydrate with no lasting damage when water is available. It doesn't mean that they need less water overall, just that they're better able to tolerate periods where it's in short supply. Then there's their hooves. Equid hooves contain microscopic keratin tubules, which function to draw water up from the environment. In donkeys, the tubules are more open than in horses, which is a useful adaptation for life in arid lands; however, in temperate climates they can become waterlogged if they are left standing in wet fields for long periods, which can lead to serious hoof problems. Wet weather presents another problem for donkeys: their single-layer coat lacks the oils that help to waterproof horse coats and the hairs don't increase in weight over winter. In contrast, native pony coats can increase by 200 per cent to deal with the colder months.

Clearly, horses and donkeys are not the same when it comes to their anatomy and physiology, and there are clear behavioural differences too. Wild horses are herd animals, adapted to life on the plains where their main food is widely abundant. They don't need to hold territories, they simply range as a group over the wider landscape. Wild asses evolved in an environment in which food and water are patchily distributed, and one that may lack the predator visibility afforded to horses on the open plains. Being a herd animal here makes little sense; instead, the only permanent social grouping is a jenny with her foals. This environment also creates the conditions for competition and territoriality. Donkey jacks improve their chances of finding a mate by holding a territory, usually around a water source, and jennies also show territorial behaviour to guard food and water.

A related point is that donkeys and horses sit at either end of the 'fight or flight' scale. When horses perceive a threat, their instinct is nearly always to run away. As potential prey, they are competing with the rest of their group to avoid being the slowest, meaning there has been a strong selection pressure for speed and leg power. Donkeys, on the other hand, tend not to run when they are threatened or frightened, instead digging in their heels and considering their options. Fleeing may not always be the prudent option, and if it's not, a donkey will stand its ground and fight. Not only is their fight instinct more easily switched on than in horses, their anatomy suits it too. Upright hooves and closely positioned legs evolved for moving around over rocky terrain, not for sprinting, and they can inflict serious damage with those big teeth and powerful hindleg kicks. In fact, the combination of strong territoriality and fight response means that donkeys can be excellent guard animals: ranch owners from the US to Australia use them to protect livestock against foxes, coyotes and wild dogs.

The tendency of donkeys to freeze, as well as their stoicism and strong sense of self-preservation, has led to their prevalent characterisation as stubborn animals. It's a commonly held belief: a YouGov poll of more than 2,000 members of the British public found that more than half of them held this opinion. For Burden, it's a frustration: 'I would love it if the public would re-evaluate the long-held perceptions that donkeys are stubborn and stupid, both of which are wrong.' Instead, she says, donkeys are 'safe and thoughtful' – traits that we have exploited to help us travel through dangerous environments and to transport our most previous wares. Unfortunately, these same characteristics have enabled a level of cruelty and abuse that exceeds that of more expressive animals, because their lack of outward expression can be mistaken for an absence of pain and fear.

It comes back to this natural tendency to think about donkeys in horse terms, as Ben Hart, animal behaviourist at the Donkey Sanctuary, explains: 'I do think that donkeys are mistreated and abused because they're measured on the horse body-language scale. People are used to seeing the very expressive nature of horses both in pain and fear. So that's the standard that people then measure donkey behaviour against.'

In fact, there is no evidence that donkeys feel any less pain than other equids. One study using an electroencephalogram (EEG) to look at brain activity in donkeys and ponies found that castration elicited a similar or even greater response in an area of the brain associated with the processing of pain. Comparing donkeys against horse standards is not appropriate for many aspects of their biology and expressiveness is no different. The consensus among equine behaviour experts today is that donkey behaviour is simply a lot more subtle.

The YouGov study also found that three-quarters of respondents did not think of donkeys as intelligent and yet there are also many people armed with stories that appear to show otherwise. A cursory look on YouTube brings up numerous examples of donkeys opening gates, including one highly viewed clip from a donkey sanctuary in Italy that shows an individual lifting a horizontal fence pole off its rests and dropping it on the floor, before it and a few other donkeys squeeze out. There are stories about donkeys being led to market in central Addis Ababa in Ethiopia and returning along the busy main road on their own, because they have learned the route and are calm enough to cope with traffic. Or of donkeys in a South American village being trained to descend to the nearest well, be loaded with water and return to the village, unsupervised, four times a day.

Nonetheless, it's hard to conclude much about donkey intelligence from these anecdotes. We need to consult the scientific literature, but here we encounter a problem: donkey cognition is practically unstudied. It's remarkable, really, when you consider the length of our alliance with these animals and the considerable research effort directed into studying the minds of other animal groups. We seem to have just accepted that donkeys lack intelligence, without any scientific support.

Leanne Proops, comparative psychologist at the University of Portsmouth, was studying for her master's degree as the field of canine cognition was starting to become established and she noticed the lack of similar studies in equids. She says: 'It seemed to me that many of their cognitive skills lay in social cognition, yet any work that was being done at the time primarily focused on basic non-social learning abilities or performance-related studies. There were no studies of donkey cognition, and there still remain very few, so equid cognition seemed like an interesting area to pursue.'

The situation is better for horses, which Proops attributes to practical factors (there are many more horses than donkeys in the countries where most cognition research is conducted), but much is still unknown about these animals. The reason, as for dogs, is that same bias against the use of domesticated animals for behavioural research. You may recall that Miklósi and Horowitz had received dismissive feedback early in their careers when they started to pursue research interests in dogs (see Chapter 3), and it was the same for domesticated equids. 'I'd say, potentially worse,' says Proops, 'because studies on dog cognition became "acceptable" before those on horses. I think there still remains an idea that equids are not very intelligent so are not a good model for cognition research.' Today, that view has been completely overturned for canine cognition; we can only wait and see whether the same will happen for equids too.

It's generally 'pretty easy' to carry out research with domestic equids, Proops tells me. Animals from sanctuaries and equestrian centres tend to be well trained and well looked after, making them ideal research subjects. In one study, Proops and colleagues evaluated the learning ability of donkeys, horses and mules. Each animal was provided with a choice of two buckets, one of which contained food. Next to each bucket was a card with a different symbol on it, with one symbol consistently put next to the food and the other to the empty bucket. It's a test of whether and how quickly the animals can learn an association: could they learn that the symbol was the thing that predicted food, and could they remember this over subsequent trials? If they 'got it' and were reliably picking the food bucket, what would happen if those patterned cards were replaced with a new pair – would the animals need the same number of trials to learn the discrimination or would they get there faster because they were 'tuned in' to the general concept of the experiment? This is known as forming a 'learning set' and the whole process is a standard method for studies of animal learning, as Proops explains. 'If an animal can learn about learning, then they become quicker at remembering each set because they learn the rule of the task.'

All animals learned to pick the food bucket at above-chance levels for at least one pair of cards, but this was as far as the donkeys got. The ponies did a bit better, with four of the six tested animals learning two pairs, although they didn't learn the first pair any faster than the donkeys. The mules were the stars of the show, with five of the six tested animals learning three stimulus pairs. What's more, mules got there quicker on each successive pair, suggesting they may have acquired a learning set after just one or two stimulus pairs. Previous studies with horses found that they did eventually seem to understand the principle of the tasks, but only after six stimulus pairs were presented; this suggests that horses may require longer to 'learn how to learn'. Donkeys were bottom of the class and it is not yet known whether they can form learning sets at all.

It's important to note that all animals tested were rescue animals with different previous experiences; these may have led to differences in performance, particularly in their motivation to engage with the task. Still, the significantly better performance of the mules compared with their parent species suggests that the hybrid vigour apparent in their physical traits may extend to their cognition too.

A subsequent study looked at detour behaviour in equids and dogs. In these tests, the animal can see a reward, but a barrier prevents them from taking the most direct route to it – they must detour around it to get to the food. The animals had four trials in which the detour was on one side of a barrier (A trials) and then four trials in which the detour was on the other (B trials). A key question is whether, once the animals get the hang of the A trials, they continue to go to that side on the B trials, even though it's no longer correct. This is called the 'A-not-B error' and it provides an idea of the animals' perseverance. The more that an animal perseveres with a learned behaviour when it is no longer useful, the less cognitively flexible they are and, as we already considered, cognitive flexibility is a hallmark of intelligence.

All the animals quickly got the hang of the A trials and navigated the simple barrier. On their first B trial, however, dogs and horses showed perseveration by continuing to approach the A side, while mules and donkeys picked a side at random. After that, donkeys were straight to the B side from their second trial and thereafter. Mules and horses got it from their third trial and dogs from their fourth. Donkeys were a little slower than mules on each trial, but Proops says: 'I have heard the saying "donkey time" and donkeys tend to work at their own speed! So, I don't think the difference in speed necessarily reflects a difference in cognitive ability, it is a species difference.' Given the reputation of both donkeys and mules for stubborn inflexibility, this test was revealing.

Paolo Baragli, researcher at the University of Pisa, used a similar set-up to investigate detour behaviour and short-term memory with Amiata donkeys, which are an endangered Tuscan breed. The animals watched food being hidden behind one of two barriers and were then held for either 10 or 30 seconds before being allowed to explore the arena. Both groups chose to detour around the correct barrier at above-chance levels, showing that they can remember the location of a reward for up to 30 seconds. This suggests that donkeys' short-term memories are no different to those of other animals.

The paucity of scientific studies on donkey cognition makes it difficult to form a conclusion about donkey minds, so I turn to the experts – are donkeys intelligent? 'This is the donkey's best kept secret!' Burden tells me. 'Donkeys are extremely thoughtful creatures, they take their time to consider the world around them, they do not act in haste and rarely act in the "wrong" way – wrong to us at least! To the donkey it is the "right" way of course.' She tells me that the animals are accurate and easy to train, assuming the right methods are used, can learn and retain knowledge, and are capable problem-solvers. These are all hallmarks of intelligence. Hart agrees: 'If you get the training right, then their speed of learning is very similar to species of dolphins and dogs, which are considered to be intelligent learners. I don't see much difference between horses, donkeys and mules, and those two.'

Fact or fiction?

The perception of donkeys as low-value domesticated animals is longstanding. Horses are bigger and faster, and their association with nobility and war has enhanced both their image and price. Cattle ownership conveys wealth and social status in many regions, and in numerous places where donkeys were traditionally used for transport, motorbikes have since taken over. Nonetheless, for hundreds of millions of people worldwide, donkeys provide vital support as beasts of burden; a role that they have performed for us for more than 6,000 years. Clearly, they are not model organisms for cognition research. The few studies that have tested their smarts suggest that they're neither the best nor the worst when it comes to learning and problem-solving. They are, however, more flexible than we (and Aesop) have given them credit.

Considerably more research has been done with horses; nonetheless, studies of horse cognition were also set back for decades, a combination of the anti-domestication stance among psychologists and negative perceptions of horse cognition following the famous, sad story of Clever Hans. Hans was a horse who, together with his owner Wilhelm von Osten, achieved widespread fame in the late 1800s and early 1900s because of his seemingly phenomenal levels of intelligence. He could count the number of people in an audience, read the clock and recognise playing cards. His most impressive ability was solving mathematical problems, which von Osten read out and Hans responded to by tapping out the correct answer on a wooden board. Hans and von Osten made international headlines and spectators would flock to shows, enthralled by the performance. Early attempts to uncover what sceptics supposed was an elaborate ruse by von Osten failed: Hans responded just as well if von Osten was not around and the questions were asked by a different person. Most people, including the Berlin Academy of Sciences, were convinced – Hans was a wonder horse! But psychologist Oskar Pfungst remained sceptical and he set out to discover what was really going on. In a devastating report of his investigations published in 1907, Pfungst convincingly debunked von Osten's claims. He found that if the questioner and the audience did not know the answer to the question, Hans failed. And if Hans could not see the questioner or the audience, he also failed. Pfungst had found that Hans was not counting in his head, he was simply responding to unintentional behavioural cues provided by von Osten and the audience. For example, people showed very subtle changes in body posture when Hans got to the correct answer, which told him when to stop tapping. When he stopped at the correct point, he received a treat, so he quickly learned to associate those behavioural cues with food.

The legacy of Clever Hans lives on: it is a story that is wheeled out with a wagging finger to all students of animal behaviour, a stark warning of the perils of not controlling for experimenter cues. As a result, animal cognition tests are now conducted 'blind', meaning that the experimenter is either out of sight of the animal or, if the animal is faced with a choice of options, unaware which one is correct. The sad outcome is the effect that the exposé had: von Osten, who was thoroughly convinced of Hans' intelligence, was devastated and died two years later. Hans was passed between owners and later drafted as a military horse – there is no further mention of him after 1916.

The reaction from the scientific community at the time was largely one of relief, because it confirmed what they had long believed to be true: horses, as domesticated animals, could not be intelligent. In fact, over the past 20 years, equine cognition has become a dedicated research topic, with specialist groups being established around the world. As Proops had noted earlier, the first studies investigated fundamental learning and perceptual abilities, finding that horses can quickly and reliably learn concepts (such as picking the largest object in an array) and remember what they've learned up to 10 years later. But, as more attention was paid to the evolutionary and ecological bases for cognition, researchers started to think more about the forces of sociality and domestication in shaping horse cognition. It's revealing some remarkable abilities.

Horses are fantastically adapted for social life, and can recognise and communicate with their herd mates through vocalisations, body language and expressions. In fact, horses share a surprising number of similar facial movements with humans and other primates. This conclusion was reached by a team who, in 2015, published their detailed 'manual' for identifying and coding horse facial movements, called the Equine Facial Coding System (EquiFACS).* They identified

---

*EquiFACS was the sixth animal FACS to be developed, after ChimpFACS, MaqFACS, GibbonFACS, OrangFACS, and DogFACS. CatFACS was added in 2017. All seven of these are adapted from the original human FACS, first published by Paul Ekman and Wallace Friesen in 1978.

17 different facial expressions in horses, which is more than has been found for dogs and chimpanzees, but less than domesticated cats. Experimental evidence has shown that horse expressions do influence the behaviour of their herdmates. For example, they're more likely to approach a photograph of a horse with a relaxed facial expression and avoid a photograph of a horse with an aggressive expression, and when they view the latter their heart rates increase more quickly than when the 'model' is relaxed.

The social cognitive skills used by horses to communicate with other horses are even more interesting when applied to the human–horse relationship. That's because a more positive way to look at the saga of Clever Hans is to recognise that Hans was fantastically skilled at reading subtle human expressions and body language. As we've seen already with dogs, the extent to which domesticated animals respond to human expressions or attention is worthy of study, because it provides a window into how these animals have evolved to live alongside us. Dogs may be a special case, given the length of their domesticated status and the fact that they typically live with us, separated from others in their species. Domesticated livestock, on the other hand, are kept in groups and for the purposes of production; they do not have the same close relationship with their owners. Horses sit somewhere in the middle, and therefore offer an interesting species in which to investigate their use of human cues. Equids haven't been domesticated for as long as dogs, but given the fact that silver foxes, selected over less than 60 years for their single measure of tameness, have demonstrated attention to human gaze (more in the next chapter), it seems possible that the 6,000 or so years of equid domestication have honed certain behavioural traits too.

Several studies have shown that horses recognise individual people and can remember their owners' faces after several months apart. Horses are also finely tuned to human expressions. They prefer to approach people who exhibit a submissive compared with dominant posture, even when they've previously experienced receiving treats from both. They can also remember and respond differently to photographs of happy or angry human faces. Horses, like several other mammalian species, demonstrate a peculiar and useful tendency for how they view certain stimuli. Negative or potentially threatening stimuli tend to be preferentially processed by the right hemisphere and this is indicated by a left-eye gaze bias. Conversely, positive, social stimuli are preferentially processed by the left hemisphere, indicated by a right-eye gaze bias. These lateralised responses provide a handy window into how the animal is experiencing the world. And because their eyes are placed on either side of their head, as is the norm for prey animals, horses must turn their heads to look with a particular eye, making it easy to record which they are using. In one of Proops' studies, horses were shown a photograph of a person either looking happy or angry and, a few hours later, the same person then visited the horse (this time wearing a neutral expression). Critically, the person did not know which photograph the horse had seen, ruling out any possible Clever Hans effects. Horses that had seen the angry photo spent more than four times longer looking at the real-life person with their left eye than those that saw the happy photo. Conversely, horses that had seen the happy photo spent 10 times longer looking at the person through their right eye than those that saw the angry photo. It shows that horses have a memory not just for human faces, but for the expression on the face too.

Sensitivity to human expression and emotional states is impressive, but horses seem to tune in to other human cues too. Research has shown they can make use of several pointing gestures to find hidden food, and they differentiate between people that are either facing a food bucket or have their back turned. They also respond differently to people who were either unwilling to share food with them, or willing but unable. Just like primates, horses show more interest in people who were willing but unable to share, raising the possibility that they may be sensitive to human intentionality.

So far, we've considered how horses make use of our communicative cues. But there's fascinating evidence that horses can communicate with us too. One study found that if horses are set an impossible task (food is placed out of their reach but in reach of a human caretaker) they will attempt to get human help. First, they try to get the caretaker's attention, by staring at them or walking over and nuzzling them. With eye contact made, they start to alternate their gaze from the person's eyes to the food and back again, just as dogs do when they're trying to communicate with their owners. This progresses to a more pronounced head jerk if the person doesn't respond. If the caretaker had not seen the placement of the food, the horse's signalling towards it was more intense, suggesting that horses may know something about the caretaker's own knowledge state.

Another study showed that horses could be trained to communicate their blanket preferences with their trainers. The animals first learned the meanings of three different symbols painted on white wooden boards: a black vertical stripe meant 'take blanket off', a black horizontal stripe meant 'put blanket on' and a plain board meant 'no change'. The 22 horses that completed the training then had the opportunity to select a symbol, which resulted in either their blanket being removed, a blanket being put on or no change. The results were clear. On a warm, sunny day, the 10 already blanketed individuals chose the 'take blanket off' symbol, while the 12 unblanketed individuals chose 'no change'. Conversely, when tested on a cold, wet day, the 10 already blanketed individuals chose the 'no change' symbol, while 10 of the 12 unblanketed individuals chose 'put blanket on'. The researchers noted that some individuals also tried to attract the trainer's attention when it wasn't their turn to be tested. On these occasions, the trainer provided them with the symbols and all chose the 'take blanket off' symbol. All the horses were sweaty underneath. It's a fascinating finding that needs replication. Nonetheless, the ability to use symbols to communicate preferences with humans has been demonstrated with some other animals, including apes, dogs and dolphins, and suggests that we might not be as far off devising a way to talk to animals after all.

We're learning that far from being dumb domesticated animals, horses possess sophisticated social cognitive abilities – for interacting with others in their group and with humans. Yet we're still a long way off understanding the extent of their intelligence; a fact that isn't just down to the number of studies conducted. Here's Hart again: 'Do we know how intelligent horses can be? Not really, because most horses aren't allowed to be intelligent. People don't want thinking horses.' He explains that if horses are brought up without being allowed to express their natural behaviour, solve problems or be confident, then how they perform in studies may not accurately represent their capabilities. It's a view that is echoed throughout the equine literature – get their training right, allow them to be social and have the freedom to express themselves, and horses might surprise us with what they can do.

The same is likely true of donkeys – the animals that have been tested to date have all had different upbringings and endured different kinds of hardship. What could we learn from animals that are allowed to develop 'normally'? Might donkeys, having a similar domestication history to horses, also be sensitive to human attentional and emotional states? 'Of course,' says Proops. 'I haven't explicitly looked, but I wouldn't be surprised if they were.' Burden agrees: 'Equally exciting to me is their emotional intelligence; donkeys are very understated in the way that they express this, but I can only say that they appear to have an ability to emotionally connect with people, and to react to and reflect our own emotional situations. I'm sure this is one of the reasons that they have been a longstanding companion of humankind.'

It's possible that in the coming years, our perception of donkey intelligence will change – the result of more studies that are tailored to their natural behaviour. We may learn that underneath that placid stoicism donkeys possess sophisticated social cognitive abilities. But even if we don't, perhaps that shouldn't matter. Donkeys have shown themselves to be patient, hard-working, loyal and dependable. Our histories have been intertwined for thousands of years: from the ancient societies that they helped to develop, trade and prosper, to the hundreds of millions of livelihoods that they support in the present day. The subtleties of their behaviour have led to misinterpretation for centuries, if not millennia, if Aesop's fables accurately depict how they were considered in ancient Greece. But it's time for these preconceptions to catch up with the facts. Far more is now known about donkey ecology and evolution, and with this it is possible to appreciate the marvellous adaptations that make donkeys uniquely adapted animals, not poor man's horses.

How, then, should we think about donkeys: are they smartasses or dumbasses? 'How about just asses?' Proops suggests. 'All species have particular cognitive skills that they need to survive. I think equids have excellent memories and social awareness, and I'm sure we still have lots more to learn about their skills.'

CHAPTER FIVE

The Fox and the Crow

A Crow, having stolen a bit of meat, perched in a tree and held it in her beak. A Fox, seeing this, longed to possess the meat himself, and by a wily stratagem succeeded. 'How handsome is the Crow,' he exclaimed, 'in the beauty of her shape and in the fairness of her complexion! Oh, if her voice were only equal to her beauty, she would deservedly be considered the Queen of Birds!' This he said deceitfully; but the Crow, anxious to refute the reflection cast upon her voice, set up a loud caw and dropped the flesh. The Fox quickly picked it up, and thus addressed the Crow: 'My good Crow, your voice is right enough, but your wit is wanting.'

I discovered foxes on 9 January 2018. I'm a grown woman so I know that sounds odd, because foxes are everywhere and everyone knows about them. What I mean is that on that January morning I first *noticed* foxes. My partner, John, had bought me a camera trap and I eagerly set it up night after night in our suburban Oxford garden, waking each day with the excited anticipation of a child on Christmas morning. Several cats' bums and a soggy mouse later, there was a tantalising glimpse of fox. It was a dark, underexposed photo (I was still learning) showing a delicately poised back leg and tail, but it was unmistakeable. That telltale bushy brush we first learn about as children, snapped by my camera at 3.25 a.m. that frosty winter night.

A fox in our garden! I became interested, reading up on urban foxes and their habits. As a student in Bristol, I had once gone out radio-tracking foxes overnight, helping[*] a researcher in the mammal behaviour group. I remember the thrill of tracking these animals through the deserted streets and graveyards. After that, my interests started to settle more on bird behaviour and I barely gave foxes another thought. Now, though, foxes had my attention – and they got even more of it over the next few months. A couple of weeks later, I was woken by the piercing, shrieking wail of a vixen on heat. Sitting up in bed, my initial confusion (and, let's face it, terror) gave way to delight when I realised what the noise was. And in early June I was woken again. This time, a high pitched, almost metallic, chinking noise was being repeated, over and over. It was like nothing I had ever heard and it was in the street. It took me a while to notice, as I peered through the blinds and my eyes adjusted to the dawn light, but there! Movement. A fox crossed the road, into our neighbour's drive then up and over the wall into the next along. And there! Over the road. A little fox cub, half-hidden in a privet bush,

---

[*]Helping is pushing it. I was 19 and had shrugged off the advice about sleeping during the day, meaning that at one point I lay down in the road and almost passed out with exhaustion.

stepped forward from the bush and then retreated. The first adult fox was back in the street and where she'd been, in the drive two doors down, another cub could be seen. This was the noise-maker, while the one across the street stayed more or less sheltered in the privet and a third scurried into a different driveway. I watched for what seemed like hours, entranced by the behaviour of this new family. The vixen was ever active, hurrying between them, over the road and from drive to drive. I was tired at work the next day but I didn't care; I felt the greatest privilege to have had that glimpse into the lives of foxes. *My foxes.*

I followed the development of this family over the next few months. There were four cubs and they were boisterous! That summer, the UK sweltered through a prolonged heatwave and I was putting out dishes of clean water every day. The reward was literally hundreds of video clips of animals in our garden, which I eagerly scrolled through over breakfast each morning. The fox family were regulars and provided the majority of the footage, the cubs rushing around chasing each other, jumping on ragged-looking adults, knocking over plant pots (I didn't mind) and playing with various objects they had brought into our garden. One of them found some stray garden twine coming out of our shed and tore around the garden with it. I was enthralled. There were also hedgehogs that, to my delight, became regulars and numerous birds that used the water to drink and bathe in during the day. On the morning of 2 July, I was thrilled to see that one of the 'teenagers' had stopped in for a drink, barely an hour before I got up. It was a beautiful creature and it couldn't have posed better. Sitting side on, it drank from the dish then, with a lick of the lips, turned and looked directly at the camera. Its pale eyes were changing from blue to amber and its ears were too large for its head. It had stocky legs, a chunky little body, a flank dusted with white. I fell in love with this little fox.

Of course, the cubs grew up. But one young female*
stayed on, and I continued to see her through autumn and
winter. Then, one morning in early December, as I was
getting ready to leave for work, I heard frenetic cawing in
the garden. I looked out and there was the fox, bold as brass,
padding along the wall with next door's garden. From the
small trees next to the wall, a pair of crows were scolding
her and she paused to look up at them, poised, a picture of
delicate balance. She then did something completely
unexpected: she raised herself and stood on her hind legs,
front paws resting in the branches, head tilted up towards
one of the crows. Aesop's fable sprang to mind, with the
tantalising thought of what was being communicated
between these creatures.

★ ★ ★

The numerous Aesopic fables that feature foxes usually
portray them in the same way: the fox is the cunning, quick-
witted animal that tricks another creature to get ahead or get
out of trouble. 'The Fox and the Crow' is one such example;
'The Fox and the Goat' another, where a fox that is stuck in
a well lures an old goat to join him and then tricks the goat
into helping him escape while the goat remains trapped. In
'The Sick Lion', the fox has the wits to notice that of the
numerous animal prints going into the lion's cave, not a single
one comes back out; he therefore declines the lion's invitation
to go inside. Perhaps most revealing, 'The Fox and the
Leopard', in which the two animals argue over who is most

---

*Obviously, I have no expertise in sexing foxes. But dog foxes are
usually bigger than vixens and tend to have larger, broader skulls. And
female cubs are more likely to stick around the territory where they
were born, while males tend to disperse.

beautiful. The leopard stakes his claim on his patterned fur, proudly showing the fox the various decorative spots. The fox responds by saying, 'And how much more beautiful than you am I, who am decorated, not in body, but in mind.'

Aesop's fables clearly played, and continue to play, an important role in cementing the fox's reputation for cunning. Aelian, the second century AD Roman author, called the fox a 'crafty creature', a 'rascal' and a 'master of trickery' when it comes to catching their dinner, talking about foxes 'plotting against hedgehogs' or sticking their tails in the air as decoys to attract bustards. Aelian also described how the people of Thrace used the fox as an indicator of whether to cross a frozen river: the fox was said to put its ear to the ice and listen for the sound of flow beneath – if it didn't cross, the people didn't either.

These same character traits have found their way into myths and folklore, such as the literary cycle of Reynard the fox, a series of popular, satirising tales that arose in medieval Europe. So popular were the tales of Reynard in France that by the middle of the thirteenth century *renard* had replaced the word for fox, *goupil*, in common speak.

Just like the wolf, medieval Christians considered the fox to be a familiar of the devil and bestiary descriptions allude to its cunning nature:

The word vulpis, fox, is, so to say, volupis. For it is fleet-footed and never runs in a straight line but twists and turns. It is a clever, crafty animal. When it is hungry and can find nothing to eat, it rolls itself in red earth so that it seems to be stained with blood, lies on the ground and holds it breath, so that it seems scarcely alive. When birds see that it is not breathing, that it is flecked with blood and that its tongue is sticking out of its mouth, they think that it is dead and descend to perch on it. Thus it seizes them and devours them. The Devil is of a similar nature.

Crafty, cunning, sly. These are character traits that have spanned the past two millennia, adjectives that slot together with 'fox' so smoothly that you need not even pause for thought. Modern language reflects this too: when someone gets ahead through nefarious means we might say, 'You sly fox!' If we are on the wrong end of such a person, we might lament being 'outfoxed'.* Perhaps you remember Blackadder's response to Baldrick proclaiming that he has 'a cunning plan'?

Fox tricksters aren't confined to the Western world, either. Several Asian cultures feature foxes in their folklore, such as the 'kitsune' in Japan, 'huli jing' in China and 'kumiho' in Korea. There are certain differences between the countries, but the common feature is of a fox that shapeshifts and tricks people. The Japanese kitsune, for example, is a fox spirit that takes the form of a beautiful young woman and seduces men to drain them of their life force.

Countless other tales exist. There are foxes playing dead to lure creatures near enough to capture, foxes leading predators on merry chases across the countryside, foxes deceiving other animals so they can steal food. It's so easy to jump to the conclusion that all these anecdotes must prove something, that all these observers cannot be wrong. But each person that has a tale of fox trickery also grew up with stories, possibly Aesop's fables, which portrayed foxes in this way. How simple it is for the human brain to connect the dots when unconscious bias is at play. The trickier question is whether our intuitive beliefs are backed up by facts. What does modern science have to say about the fox?

★ ★ ★

---

* There are other uses too. Being called 'foxy' is a compliment for either sex, meaning attractive or sexy. The term 'silver fox' has been used for decades to describe a handsome man with greying or white hair. These days, 'vixen' is also used to describe an attractive woman, but its earlier alternative meaning was of a spiteful, quarrelsome, shrewish woman.

Foxes belong to the *Canidae*, just like wolves and dogs. Modern canids are thought to have emerged some 12–10 million years ago and in the region of 6 million years ago they diverged into two tribes,* with the 'wolf-like' species in one (the *Canini*), and the 'fox-like' species in the other (the *Vulpini*). Within the latter, the 'true' foxes belong to the genus *Vulpes*, including the red fox, *Vulpes vulpes*, our main protagonist in these stories. Because *Vulpes* essentially means 'fox', that makes the red fox the quintessential fox. They are the largest members of the genus, yet are by no means large animals, with adult head and body length ranging from 45–90cm (excluding the tail, which can add 30–55cm) and body weight from 3–14kg. It can be surprising to learn of their diminutive size, given the size of the personality that accompanies the red fox in all stories.

According to the Canid Action Plan, developed by the IUCN/SSC's Canid Specialist Group in 2004, the red fox has an estimated global distribution in the region of 70 million km², giving it the widest geographical range of any carnivore. The only canid present on five continents, red foxes are found throughout the northern hemisphere (excluding Iceland, the Arctic islands, some parts of Siberia and extreme deserts) and Australia too, after being introduced there in the 1830s. In the past 170 years, the fox is thought to have expanded its range to nine new countries. This translates into a spectacular diversity of habitats: foxes are found everywhere from ice fields to deserts, high mountains to swamps, rainforests to cities. And while this immense geographical distribution means that there are some evolved differences between populations (with about 43 different subspecies of red fox recognised), the fox that hunts lemmings in the frozen Siberian tundra is biologically the same as the fox that digs for

---

*A tribe is an intermediate classification between the family and genus level. It's just another way of grouping the more related members of the family.

earthworms in the UK and the fox that harvests locusts on the edge of the Sahara Desert. Unsurprisingly, the red fox is classified as Least Concern by the IUCN. It's one of the few animals that is doing well out of urbanisation, whether we like it or not.

Foxes are predators, but they have their enemies too – golden eagles and Eurasian lynx will hunt foxes and their cubs, while grey wolves, coyotes and dingoes will kill foxes in clashes over territory and food. In the presence of these literal 'top dogs', foxes adjust their behaviour, showing more vigilance in places where there's a high chance of bumping into them, such as water sources in the Australian desert. Still, foxes don't defer easily: one study found that they readily added their own scent over the top of dingo scent marks, which is tempting to interpret as a vulpine two-fingered salute to their bigger rivals. So far, this doesn't contradict Aesop's depiction, but let's go a bit deeper into what we know of fox behaviour.

In 1948, a young zoologist called Günter Tembrock was given a red fox to observe. It launched his career. That same year, he established a research centre for animal psychology in the Zoological Institute of Humboldt University Berlin, which was being rebuilt after the war. It was the first such institute to be founded in Germany and what followed was 20 years of research into the behaviour of red foxes, largely conducted in the 'fox room' of the Museum of Natural History in Berlin. Tembrock was heavily influenced by the work of Konrad Lorenz and Oskar Heinroth, pioneers in the then-emerging study of animal behaviour called ethology. He adopted an ethological approach to his studies of the fox, documenting every aspect of their' lives in order to understand their natural behaviour patterns. Tembrock used drawings, photographs, films and (later) tape recordings, distilling it into his classic study on red foxes, '*Zur ethologie des rotfuchses*' ('On the ethology of the red fox'), a 243-page monograph that was published in 1957. Detailed observations, paired with black-and-white photographs, show numerous examples of instinctive behaviours in captive foxes; for example, going through the motions of burying food in a tiled room, even though it was unable to dig a hole.

Wildlife biologist David Henry also took an ethological approach to his studies of red foxes, wanting to enter the fox's umwelt, just as Horowitz had advocated with her dog research (see Chapter 3). His 14 years studying the animal in Prince Albert National Park in Saskatchewan, Canada, led him to term it the 'catlike canine', a very apt description. There's the measured stealth, silence and sure-footedness that comes with being a master stalker, the habit of laying up in a sheltered, sunny spot during the day and hunting at night, the fact that foxes, like cats but unlike any other canine, have vertically slit pupils and can climb trees. In Henry's words: 'Aesthetics and adaptation are one in the red fox – its most exquisite features are also some of its most important tools for survival.' The similarity is not because they're closely related, because while both sit in the broad overall group of *Carnivora*, cats are members of the family *Felidae*, which last shared a common ancestor with canids in the region of 50 million years ago. Instead, cats and foxes show a beautiful example of evolutionary convergence. Both evolved to exploit the same ecological niche of hunting rodents and other small mammals. Since their main prey is nocturnal, they are too, and that requires adaptations for light sensitivity, such as the vertically slit pupils. Rodents are extremely sensitive to noise, meaning that to have any chance of catching them, their predators must be stealth specialists. Foxes are just about as catlike as they can get, within the constraints of being a dog.

Today, different aspects of fox biology are being studied by research groups around the world, including Australia, Japan, Mongolia, Russia, the US and several European countries. Much of the work is applied: in Australia and Tasmania, for example, there's a strong and necessary focus on evaluating how much damage the introduced red fox is inflicting on native species and its possible management. Other key research themes include monitoring the spread of pathogens, including rabies, fox tapeworm and sarcoptic mange.

At Oxford University, Professor David Macdonald is the head of the Wildlife Conservation Research Unit (WildCRU). Today, the group's research covers species and topics across the spectrum of global conservation biology, including clouded leopards in Borneo, Himalayan wolves, Andean bears, African lions and, closer to home, a 30-year dataset on the ecology and behaviour of European badgers. But Macdonald's research career started with foxes, seeded as a child when he discovered fox pawprints in the bunker of a local golf course. Macdonald began his fox studies in 1972, part of Niko Tinbergen's world-leading Animal Behaviour Research Group at Oxford University and supervised by Hans Kruuk, a specialist in carnivore behaviour. At the time, naturalists saw the red fox as an antisocial loner, the total opposite of its more sociable cousin the wolf. Wolves travelled in packs, foxes travelled alone – and this difference was traditionally explained by the size of their respective prey. Wolves, jackals and wild dogs bring down large herbivores in groups and even coyotes, which are usually solitary animals, come together to hunt larger animals such as deer. But have you ever heard of the pack of foxes that joined forces to bring down a vole? Exactly. Foxes are unusual canids because they hunt small animals that suffice as a meal for one. They wouldn't benefit from cooperative hunting and it would probably be a hindrance anyway. Rodents are dinner for a lot of animals, so they've evolved some fantastically fine-tuned predator–detection systems. They are highly strung little animals, always alert, their senses attuned to the slightest movement, noise or smell. A solitary fox must channel all its stealth and lightness of foot to catch a rodent, so a skulk of foxes stands no chance! Wolves took a different route; openly pursuing their quarry, they rely not on stealth but on stamina and teamwork to capture their prey.

The theory essentially said that wolves need buddies and foxes do not, and while that was plausible, Macdonald and others had nagging doubts, having seen larger groups of adults and cubs that didn't seem to be families. For them, the red fox was the perfect animal in which to investigate flexibility in social systems.

Foxes are elusive, secretive beasts; persecuted by humans for centuries, they actively avoid our presence in the countryside (less so in urban settings). This makes tracking, identifying and studying them more than a little bit challenging. For the first half of the twentieth century, skilled fieldcraft was essential, as depicted in Adolph Murie's *Following Fox Trails* (1936). Murie, who we already encountered in Chapter 2, studied a pair of foxes in Michigan during 1934, using the traces of prints left in snow to glean information on their behaviour. Through today's lens of high-tech gadgetry, it seems incredible to think that this method could reveal much about fox habits. Murie, in contrast, wrote that tracking 'is practically equivalent to observing an animal for a long period of time under natural conditions'. For this expert in fieldcraft, the trails were akin to a sign language that just needed some interpretation to be understood.

Fieldcraft is tough, even for those with Murie's skill, and it provides only brief insights into fox life. Fortunately, from the middle of the century great advances were being made in the technology of radio-tracking, something that would revolutionise the study of wild animals. Of course, to fit a fox with a tracking device you still needed to catch it and that remained a challenge. Macdonald's early days were full of frustration and failure: the chicken-liver-baited traps that he carefully set out around an obliging farmer's fields were peed on, defecated in, dug under and climbed on by foxes, but he never caught a fox with this method. Once he and Kruuk worked out a better technique, the study was under way. Every captured fox was fitted with a collar that contained a little radio transmitter, which emitted signals at a specific frequency. A special, directional receiver enabled researchers to tune into each of these. It was a groundbreaking advance for studying fox biology because it revealed where each animal was in time and space, enabling a much more complete picture of their habits to be built up.

Macdonald became a familiar figure to many of the local people living within his study area, earning him the nickname of 'Foxman'. Night after night he was out stalking the animals; he came to recognise individuals, he learned their routines so that he could predict where he might find a fox laid up during the day and, importantly, with which other foxes that animal interacted. This was a previously unimaginable insight into the social lives of the red fox and Macdonald's three years of intensive radio-tracking thoroughly debunked the notion of the fox as an anti-social loner: he found that more than 20 adult foxes lived in just over 2km² of Oxfordshire countryside! His research showed that several vixens associated in a group, usually with a dominant dog fox patrolling the home range. The vixens' relationships were hierarchical, often comprising a dominant mother and her adult daughters. The subordinate, non-breeding vixens sometimes stepped in to help bring up cubs; for example, when the dominant female was sick, injured or had died.

The semi-rural foxes that lived in the affluent area north-west of Oxford's ring road yielded insight after insight and, like any good research project, raised more questions than were addressed. The techniques pioneered in this population were subsequently applied to comparative studies of foxes in other localities: Oxford city centre, the Cumbrian Fells, the

Dead Sea in Israel and arctic foxes in Iceland, to name a few. The city centre studies started in the late 1970s and are particularly compelling for me, given that the area of Oxford that was studied is the same one in which I live and made my own, novice observations of fox family life. The group (termed, the 'Oxford Foxlot') tracked their subjects from a converted old Hackney carriage, the glowing 'TAXI' light replaced by the word 'FOXLOT'. A hole in the ceiling accommodated the tracking apparatus and the taxi became a well-known sight, with most locals aware that this was one taxi that could not be flagged down.

More on urban foxes later. First, what research has uncovered on the basic biology of red foxes. Dietary studies have been at the forefront of much of this, both to implicate and exonerate foxes of feasting upon livestock. Foxes are carnivores, because they're canids and they possess the characteristic arrangement of teeth that defines a carnivore. And although we know more about the ecology and behaviour of urban foxes, enough is known about rural red foxes to appreciate that they can be a real problem for farmers. Foxes will predate chickens, ducks, turkeys, piglets and newborn lambs, with significant economic consequences. One study estimated that they killed 1–2 per cent of newborn lambs, and in a survey of Welsh sheep farmers in 2013, 96 per cent said that predation impacted their income. The overall annual impact of foxes on farming in the UK has been estimated at about £12 million, with more than £9 million of that from the sheep sector. The difficulty with these estimates is that it is not straightforward to work out whether foxes killed or simply scavenged the animals, meaning that perceptions of fox predation may be inflated; nonetheless, these are not insignificant losses.

On the flip side, foxes are not a farmer's worst enemy. In the 1980s, rabbits were estimated to have caused £120 million in losses to British farmers each year. Assuming that rabbits are still feasting to a similar level on farmers' crops,* then by predating rabbits and rodents, foxes may have largely offset their costs and, in some situations, may even be economically beneficial to the farming industry.

What's more, foxes are not 'hypercarnivores' like lions; they are generalist omnivores with a hugely diverse diet and no single dominant food type. They are the opposite of a koala, feeding only on eucalyptus leaves, or a Canada lynx, feeding almost exclusively on snowshoe hares. This is apparent from analyses of stomach contents or, non-invasively, from their poo. Macdonald and his student Pat Doncaster analysed nearly 2,000 fox scats collected around Oxford city in the early 1980s. They found that scavenged food made up more than one-third of the total food consumed by foxes living in Oxford, followed by earthworms, small mammals, fruits, birds and other invertebrates. Domestic stock was also taken, but at much lower levels. A finer look into what the foxes were eating revealed a staggering 81 different food types, including 14 different species of small mammal, 13 different types of fruit, and traces of plastic, glass, aluminium foil, wood and even dog hair. Foxes, clearly, are not fussy eaters – they eat whatever is abundant and they excel at finding it; they have been described as jacks of all trades and masters of most of them too. The UK-based charity, The Fox Project, states on their website that in their 29 years of operation they have rescued around 15,000 foxes and they are yet to find a starving one.

Extreme dietary flexibility enables foxes to adapt their diets through space and time. Scavenged food may make up the bulk of a city centre fox's diet, but in the Cumbrian fells, the largest proportion will be fresh rabbit, rodent or bird. In the desert regions of the Middle East, invertebrates tend to

---

*Given that the population of wild rabbits in the UK has increased since the 1980s, when it was still recovering from the catastrophic impact of myxomatosis, this is likely to be conservative.

make up a greater proportion of a fox's diet than mammals, while in the arid regions of Australia fox scats sometimes contain scavenged kangaroo remains. In Hungary foxes predate wild boar piglets, while in North America they take the risk of brazenly scavenging from puma kills. Every habitat offers something to the fox and as food availability changes through the year, they simply change too. Like the head chef of a Michelin-starred restaurant, foxes are connoisseurs of local, seasonal produce, taking advantage of whatever is in abundance at any time. That might mean feasting on berries, acorns or windfall apples in autumn, while taking advantage of ground-nesting birds' eggs or leverets in spring. For the Oxford foxes, rabbits peaked in the diet in April, birds in June and scavenged items in late winter, while fruits became the main dietary component between August and October (strawberries in August, blackberries in September and plums in October). For the most part, this mirrored the seasonally changing abundance of these foods.

What then, of the common belief that foxes kill 'for fun'; that when given half the chance, they enter a state of bloodlust that drives them to kill for sheer pleasure? It's a characterisation that is intimately tied in with our negative perceptions of foxes, but doesn't seem to be grounded in fact. It is true that foxes will go on a killing spree if they manage to gain entry to a hen house or pheasant pen. But this is better explained by their biology, not by psychotic personality traits that give them a perverse pleasure from killing. Foxes have evolved as skilled predators of specialist 'escapers', so under natural conditions a fox that can kill additional prey will be at an advantage. What they haven't evolved to deal with are human livestock enclosures, with animals that can't or don't run away. It's thought that in this specific situation the fox's natural instinct to kill is boosted into overdrive by the abnormal behaviour of the prey, resulting in them killing far more animals than they need.

Foxes sometimes experience a seasonal glut of food too, just like when our veg patch throws out a bumper crop. To prevent our courgettes or raspberries going to waste, we adapt our diet to work them into our meals, or we preserve them for later consumption. When foxes have killed more than they need to eat, they also stash the extra, making stores that they return to in leaner times. In the tundra of northern Canada, storing food is highly advantageous to manage the rapidly changing food availability. During the harsh Arctic winter, the temperature may be -50°C and deep snow makes scavenging almost impossible. Yet, in spring and summer, lemmings and massive breeding colonies of snow geese provide a feast for arctic foxes, who have been recorded carrying off up to five lemmings or three goslings at a time and storing them in the fridge-like conditions of the arctic permafrost.

When foxes hoard food they either spread it thinly in multiple sites (called 'scatter hoarding') or they group it together in larders close to their den. Larders can be substantial. With an estimated 2 million pairs of birds,* Baccalieu Island in Newfoundland, Canada, hosts the world's largest breeding colony of Leach's storm petrels. This offers a huge seasonal glut of food for foxes, who both scatter-hoard and store the birds in larders. One study reported a larder that contained nearly 400 birds! Faecal analysis during winter showed that the foxes were clearly recovering their cached birds to eat, since at least half the samples contained petrels.

Macdonald studied food hoarding in a hand-reared vixen called Niff. She lived with him in his rented room and from just four weeks old she started to hide food in every possible corner and crevice. She was fiercely protective of her caches and would move them if she noticed him watching.

---

*When the study was done in 1985, this was estimated at 3 million breeding pairs, which may have been an underestimate. The latest data come from a 2013 survey.

Macdonald began to test this experimentally when she was 10 weeks old and he had started taking her outside on a lead for walks. Beforehand he would scatter a couple of dead mice at places on the route, allowing her to cache any that she found during the walk. She was then allowed to recover them and after six months Niff had recovered an impressive 48 of 50 cached mice. To work out whether she remembered where she had stored them or whether she was just sniffing them out, Macdonald devised two more sets of trials. In one, he secretly buried another mouse within 3m of Niff's hidden one; in the other, he dug up Niff's mouse and reburied it 1m away. The results were striking. In the first task, Niff recovered almost all the mice she had buried, but only about 20 per cent of Macdonald's; in the second, she recovered only about 25 per cent of her relocated mice. Macdonald subsequently repeated these experiments with other foxes, with similar results. Foxes seem to have highly developed but very specific memories for the locations of their caches.

How does Niff's failure to recover mice that were moved just 1m away fit in with the fact that foxes, like all canids, possess a supremely good sense of smell? The sensory world of the fox is razor sharp, far more acute than our own, and it helps them to detect food and avoid detection by their predators and persecutors. As a solitary forager that communally defends territory with other family members, scent plays a key role in communication, helping the fox to keep in touch with the other members of its group. It does so by scent-marking with secretions from special glands. These complex chemical cocktails provide a rich seam of social information, allowing each fox to build up a detailed olfactory web of who passed through, when, where they were heading, as well as their reproductive and possibly even health status. Foxes have two kinds of scent gland. The supracaudal, or 'violet', gland on the top of their tail produces scent that is said to smell pleasantly like violets and constitutes a mix of chemicals derived from plant carotenoids, which are known to be important for health.* This is not the case for the scent from their anal glands, which has been described as 'rancid'. By adding their own scent over the marks of others they are in essence communicating with them, providing information about their own status. They also use scent to learn about other species. Foxes pay a lot of attention to urine, faeces or scent marks of dogs and cats, who represent predators and competitors. This means that wherever possible foxes travel into a head wind, gleaning what information they can about other individuals in the area, without betraying their own presence. In the world of the fox, tail winds are to be avoided.

Urine is also used as a foraging cue. In the 1970s in Saskatchewan, Henry observed foxes urine-marking inedible food remnants, such as bone, bird wings and dried out pieces of skin. He found that when they re-encountered these urine-marked remnants they spent less time investigating them than they did for equivalent, non-marked scraps. They also seemed to pay attention to other foxes' urine marks. Henry described how scavenging foxes may stop and investigate more than 200 spots per hour on the forest floor for several hours of the day. That's a lot of information to keep track of. He suggested that urine-marking is a kind of 'book-keeping system' for the fox, helping it to increase its foraging efficiency by signalling which items could be ignored.

Use of urine as a social signal develops in young foxes: up to the age of 5–6 months, they urinate in seemingly random places, but after this age, they start to direct it more over visually conspicuous objects such as grassy tussocks or molehills, coupled with a lot of sniffing beforehand. These sprinklings are known as 'token' urinations. From my

---

* In several species of bird, dietary carotenoids colour feathers and skin, providing a visual signal about health status.

camera-trap footage, I witnessed a couple of occasions when a fox crouched over the water bowl and sprinkled a few drops into the water. Often, this happened after the animal had drunk from the dish (she never drank from it after marking!). I still have no idea why she would do this, but I've no doubt it was intentionally aimed. Dominant foxes also token-mark other individuals, which is called 'allomarking' and is usually done to a sleeping subordinate's head or shoulder. Foxes scent mark with their faeces too, and while this is normally on top of prominent features such as molehills, Macdonald witnessed foxes defecating while standing on their front legs with their back legs raised on an object. Like a bizarre, smelly kind of handstand.

Yet, while used extensively in their social lives, smell doesn't seem to be the fox's primary sense for prey capture or recovery. Experiments by Finnish biologist Henrik Österholm in the 1960s indicated a hierarchy in the importance of different sensory inputs for foxes: under twilight and nocturnal conditions, they preferentially relied on sound over sight, and sight over smell. In fact, Österholm found that if food was buried 10cm below the surface, the animal did not locate it when further than 50cm away, which is consistent with Niff's poor performance. He concluded that foxes are much more sensitive to moving prey, which they can locate with sound or sight, referring to smell as a 'point-blank locatory organ'. His own observations in the forests supported this: he chanced upon a black grouse nest with eight eggs, just 1m from a path frequently used by foxes and 100m from a fox den. Remarkably, all the eggs hatched.

Reliance on acoustic cues makes sense when you think about the difficulties foxes face in locating prey that is superbly adapted at hiding. Small mammals may hide in the densest vegetation or, in the depths of winter, underneath thick snow, posing a considerable challenge to any predator. To overcome this, the fox has evolved a characteristic behaviour called 'mousing', which enables it to catch prey in these toughest of conditions. Imagine the animal, battling against driving winds, sub-zero temperatures and deep flurries of snow. Suddenly it stops. Head tilted, paw raised, it is completely absorbed in a patch of nondescript snow. Its ears are erect and it turns its head slowly from side to side. And then, with a flick of the tail, it jumps high into the air before plunging headfirst into the snow, like a missile locked on to its prey. The sight of a fox's back legs and bushy tail wiggling out of a snow drift makes for a comical sight, but this is a masterclass in sound localisation and an essential survival tool, enabling the capture of small rodents before they know what's hit them.

Foxes achieve this because they possess a sophisticated set of adaptations for rodent hunting. First, their hearing is both exceptional (they are said to be able to hear a mouse squeal from about 45m) and tuned to match the noises made by their prey: studies have shown they can locate sounds to within one degree in the range of 700–3,000Hz, which overlaps with the frequency of noises made by small mammals rustling in vegetation. Next, foxes can move their ears independently to help pinpoint a noise and their characteristic head tilting creates asymmetry between the ear canals, meaning that sound hits one side marginally later than the other. By comparing the sound inputs between the ears, the animal can pinpoint its exact source. Barn owls achieve the same result because their ears are physically asymmetrical: the right ear canal is higher on the head than the left. The owl evolved an anatomical adaptation; for the fox it's behavioural.

There may be more to it than just a sensitivity to noise. In a 2011 publication entitled 'Directional preference may enhance hunting accuracy in foraging foxes', a team of biologists led by Jaroslav Červený from the Czech University of Life Sciences reported some intriguing findings from an analysis of mousing bouts. They analysed almost 600 mousing jumps from 84 different foxes and found a clear, non-random tendency for foxes to position themselves towards the north-east as they prepared for each jump. What's more, when the animals were oriented in a north-easterly direction they were more successful at capturing prey, particularly in dense vegetation where they could rely less on vision. In fact, about three-quarters of all successful jumps in high-cover conditions were clustered within 20 degrees clockwise of magnetic north. It's an astonishing finding, and Červený and team first needed to check whether the animals were simply guided by other factors in their environment. They found that irrespective of the time of day, season, level of cloud cover or wind cues, the animals remained consistent in their orientation. The data suggest that foxes align their jumps to the Earth's geomagnetic system, using it as a 'range finder' to estimate distance to its prey. The angle of the Earth's magnetic field varies according to where you are on the planet, but in the northern hemisphere it tilts downwards, hitting the ground at an angle of 60–70 degrees. Červený's hypothesis is that when foxes listen to the rustling sounds made by prey, they're lining up the angle at which the sound is hitting their ears against the angle of the magnetic field. When the two slopes match, the fox knows it is a fixed distance away from the source of the sound and therefore knows how far to jump. It's a fascinating hypothesis and, if correct, would be the first example of an animal using the geomagnetic field to hunt, as opposed to navigate. Much more research needs to be done, including answering the fundamental question of just how the fox might detect magnetic fields in the first place.

★ ★ ★

We know from archaeological evidence that humans and foxes have associated for at least several millennia. For example, at the bronze age site of Can Roqueta, Barcelona, which dates to between 4,000 and 5,000 years ago, four foxes were discovered buried with humans. One of the foxes was an old animal with a broken leg, yet expert analysis of the remains showed that the fractured leg was healing and that it had been manipulated in a manner consistent with having received human attention. Other evidence for a long relationship between humans and foxes comes from the Channel Islands, off the coast of California, where a different subspecies of island fox inhabits each of six islands. It's thought that grey foxes were brought to the islands by the Native American Chumash people and subsequently evolved into the little foxes we see today. The oldest fossil remains date to 6,000 years ago and several lines of evidence indicate that the foxes were sacred to the Chumash people.

One finding suggests the presence of a relationship that goes back much further. Fox and human remains were uncovered in a grave at the pre-Natufian burial ground of 'Uyun-al-Hammam in northern Jordan, which is around 16,500 years old. More intriguingly, both sets of remains appeared to have been intentionally moved and reburied together, a practice that is thought to be an intentional, symbolic act. Based on the archaeological evidence, the research team suggested that the fox was a companion to the grave's occupant; when the grave was reopened this relationship was remembered and the fox bones were moved so that the pair would continue to be together in the afterlife.

The grave at 'Uyun-al-Hammam is around 4,000 years older than the earliest known human-dog burial, although the current consensus is that dog domestication happened at least 15,000 and potentially up to 40,000 years ago. Yet, is it possible that early humans had fox, as well as dog, companions? It didn't last if so; our ancestors must have favoured the larger, stronger, more sociable canine. But what a thought that we could be walking around with domesticated versions of foxes, instead of wolves, on our leads!

In fact, more than 60 years of research offers an intriguing insight into fox domestication. In 1959, Soviet geneticist Dmitri Belyaev began an experiment into the genetics of tameness in silver foxes; the 'farm fox domestication experiment' today ranks as one of the most important in evolutionary biology. Belyaev was familiar with Darwin's ideas of natural and artificial selection and from his time working at the Institute for Fur Breeding Animals in Moscow he was well versed in the methods of selective breeding. But just how domestication had got started in the first place was still a mystery. It was also apparent that domesticated animals shared certain traits that set them apart from their wild ancestors. What is now called 'domestication syndrome' includes floppy ears, short curly tails, juvenilised facial features, mottled coats, reduced stress hormone levels, smaller brains and relatively longer reproductive seasons. Belyaev was intrigued: cattle and dogs were domesticated for food and protection, not floppy ears and spotted coats, yet everywhere he looked he saw signs of domestication syndrome. He hypothesised that domestication must have something to do with prosociality: our ancestors would have likely chosen the calmest, least aggressive individuals in a group to keep and breed from, killing those that were aggressive or uncontrollable. Belyaev then wondered if there was a genetic link between these prosocial traits and all the other characteristics that were shared by domesticated species: was domestication syndrome simply the by-product of selection for friendliness?

Belyaev recruited Lyudmila Trut, fresh from her degree at Moscow State University, and they started to breed silver foxes (a melanistic form of the red fox that has long been trapped and bred for its fur) in the harsh Siberian conditions at the Institute of Cytology and Genetics of the Siberian Branch of the Russian Academy of Sciences in Novosibirsk.[*] They began with 198 foxes, selected from fur farms across the Soviet Union for their less aggressive and fearful behaviour towards humans.[†] In each generation, the cubs were tested on the single metric of tameness. The top 10 per cent tamest animals were selected for breeding and the process was repeated. Foxes breed every year, so since its initiation tens of thousands of silver foxes have been bred. After just six generations, foxes had gone from trying to flee from people to licking the researchers' hands, having their tummies rubbed and wagging their tails when Trut approached. It was a phenomenally fast result. Ten years into the programme, a fox was born that had floppy ears, and after that floppy ears became common and curly tails appeared too. By the fifteenth generation, levels of the stress hormone glucocorticoid were about half that of wild foxes, while levels of the 'happy hormone' serotonin had increased. Body shape changed too: foxes from the tame selection line showed shorter, rounder snouts and chunkier limbs.

Today, the tame population of silver foxes at the Institute – where Trut still leads the research – are calm, fearless and eager to interact with humans from a young age. At the other extreme, a line of aggressive foxes, selected over the same

---

[*] It was a risky time to be conducting genetics research in Stalinist Russia, after Trofim Lysenko had imposed brutal crackdowns on this area of research. Fortunately, Belyaev's fur-farm work was the perfect cover for his selection experiments, since the export of high-quality fox furs was sufficiently lucrative for the government. Nonetheless, it was better to conduct this work well away from Moscow.

[†] About 50 years of breeding at fur farms had done some of the work – the silver fox had already been subjected to selection for adaptation to the farm environment, so can't be considered a completely wild population; nonetheless, farmed foxes still exhibit fear and aggression to humans.

period for their intolerance towards people, snap and bite at the cage doors as the scientists walk past. It's a remarkable study and it continues to provide fascinating insights. The underlying mechanism has been traced to genetic mutations, which result in a reduced size and activity of the adrenal gland, an endocrine organ that is involved in the stress response and produces glucocorticoid hormones. Adrenal activity during development is thought to control friendliness, as well as the development of physical features including the size of the face and limbs, ear and tail rigidity and the amount of colour in the coat.

Behaviourally, there are also signs that the domesticated fox cubs are more in tune with humans. One study showed that fox cubs from the domesticated line, who had next to no experience with people, were as good at using human pointing and gaze direction to find food as domestic dog puppies. Critically, cubs from the control group (which has not been selected for either tameness or aggression) were not able to make use of such cues unless they had considerable experience interacting with people. Sensitivity to human communicative gestures makes sense for dogs, the result of thousands of years living side by side. The fact that fox cubs – selectively bred over less than 60 years for the single trait of tameness – also show this ability suggests that it may instead have been a by-product of selection acting on the systems that control fear and aggression.

The fox farm study shows how intense selection can cause changes in physical, behavioural and physiological traits associated with domestication. But what about selection on a more subtle scale? The age in which we live, the Anthropocene, is marked by habitat destruction and species loss, the consequences of an insatiable global drive for wealth, power and progress. As our cities continue to sprawl outwards and consume natural habitat, wildlife must either relocate and avoid being swallowed up, or become part of the human landscape. Adapting to human life is not possible for all animals, but for those that can we're already seeing evidence for adaptive change in certain traits. For example, a population of cliff swallows nesting under bridges and highway overpasses in Nebraska, USA, have become better at avoiding collisions with traffic. Ornithologists studying the population recorded a considerable decrease in the number of roadkill swallows over 30 years, and at the same time a change in wing length: recovered corpses had increasingly longer wings than the overall population average. Since shorter, rounder wings enable a more vertical take-off than longer wings, it seems that natural selection may have acted upon wing length to help them avoid collisions, adapting this population for life alongside the fast lane.

The fox provides a fantastic case study for how a wild animal has adapted to urban life. They first set up in London and other cities of southern England during the 1930s. Land was cheap and huge swathes of it were bought up and developed into the first suburbs: low-density housing zones that spread into the countryside and swallowed up many acres of fox territory. And yet, here was an opportunity for the fox: compared with city-centre terraces, the suburbs had fewer houses and larger, greener gardens. Dry, sheltered den sites could be dug under garden sheds. And there was still food waste, which could be eaten and which also attracted rats and other prey. In other words, suburbia was fantastic fox habitat. Leafy suburbs are still prime locations for urban foxes, but they will readily defend a territory in any area of a city. And it's no longer a British phenomenon: urban foxes are now recorded in cities across Europe, North America and Australia.

Stephen Harris, now retired from his position as professor at Bristol University, initiated his study of Bristol's foxes in 1977, focusing on a 1.5-km^2 patch of 1930s semi-detached housing in the north-west suburbs. Conveniently, as he reflected in a 2019 article, his house happened to be in the middle of the area and the garden next door was overgrown, making it an ideal place to get an insight into fox social life. Harris and many students and colleagues have trapped, tagged and followed foxes for more than 40 years, making this patch of urban fox territory likely the best studied of any city.

Much of the erroneous belief for foxes being solitary animals came from the fact that they hunt alone. We've already covered how this reflects their meal-for-one diet of small mammals and birds, but their nocturnal hunting activity is not fully representative of fox life – and in fact, urban foxes live in family groups and jointly defend a territory, just like their rural counterparts. Night-time might be prime time for hunting, but a lot of social interaction happens during the day in 'rendezvous sites', where the fox family group meets up. The garden next to Harris was one such site and it provided him a window into what foxes got up to throughout the day. He witnessed a lot of coming and going, with plenty of greeting and grooming behaviour, as well as play and sometimes fighting. For Harris, it was clear that foxes have complex, intricate and structured social lives, in common with other, more obviously social mammals.

The Oxford and Bristol foxes have shed light on urban fox territoriality. Phil Baker, now lecturer at Reading University, mapped out the territory sizes of Bristol's foxes and found that they could be as small as $0.2km^2$, depending on resource availability. Red fox territory size is highly variable but a rural fox may hold a territory up to $40km^2$. Not only that, but territories weren't fixed. Macdonald and Doncaster found that in Oxford, the size of the area defended by each family group remained constant, but its location didn't – like sand dunes, the honeycomb of territorial borders shifted and crept with time, reflecting disturbances inherent with the city environment.

The Oxford and Bristol foxes also provided evidence for greater fluidity in fox family lives. Like most other canids, foxes are usually considered monogamous; in reality, they're a bit sneakier. Dog foxes will trespass into other territories during the breeding season to solicit mating with unfamiliar vixens. Usually, territories of urban foxes are bordered by major arterial roads into the city that they don't need to cross, but during winter, when hormones surge, males take their chances; a trade-off that can be worth it if it results in them fathering additional cubs, but which many of them pay for with their lives. What's more, it had been assumed that subordinate vixens in a group were non-breeding 'helpers', reproductively suppressed by the dominant vixen's aggression. There is evidence that harassment by the dominant vixen can cause stress responses that prevent subordinate females from breeding. On the other hand, the Bristol team found that many subordinates do conceive, although this is highly variable between populations, dependent on how much food is available.

Advances in camera-trap technology have enabled extra insights to be made of the numerous foxes that frequent the Bristol patch, without requiring individuals to be followed. Between 2013 and 2015, Harris's student Jo Dorning set up camera traps in each of seven gardens (corresponding to a different fox territory) for 40-day periods within each season of the year. That was the straightforward part: Dorning then had to work out how to process the resulting 800,000 images. More than three-quarters were triggered by other animals so could be immediately discounted, but this still left more than 174,000 fox stills to assess. Incredibly, Dorning could use unique fox characteristics to identify individuals in 99 per cent of these images. She found that 192 different foxes had visited the gardens, and that they quickly learned which gardens offered a predictable, regular source of food and water. The data showed that female foxes were the more efficient foragers, targeting the most frequently provisioned gardens more strongly than males and more often getting there first.

In the mid-1990s, the Bristol team estimated that around 33,000 foxes lived in the towns and cities of the UK, with the highest reported density of 37 animals per square kilometre only possible in areas with superabundant food (i.e. people with gardens who regularly leave out food and water, like the leafy Bristol suburbs where Harris initiated his study). Contrast this with the density of foxes in rural Scotland, which may be more like one animal per 40 km^2. Dawn Scott, professor of mammal ecology and conservation at Keele University, has been studying urban wildlife in the UK since 2010. She and her colleagues revisited urban fox density using citizen scientists to submit survey data from 2013–15 in eight UK cities. They estimated an average of 3.63 fox family groups per square kilometre which, when mapped on to housing density, equated to an average of one fox family group every 648 houses. With an average of 3.4 foxes per social group, the research indicates that there may be something like 12 foxes per square kilometre of urban environment, which sounds like a lot until you think that there could be as many as 30 times more cats in the same area. While the overall pattern from the study was for an increase in the UK urban fox population, that's not because foxes are flooding into towns and cities unchecked; they've simply expanded their range to cover additional, previously unoccupied urban environments. In cities that foxes had not colonised 30 years ago, the research team found a fivefold increase in fox density; Newcastle, for example, was fox-free in the 1990s, but now has a fox density approaching that of London. But for most UK towns and cities, foxes have been there for decades so the population is stable; in ecological terms, it has reached its 'carrying capacity', meaning the area can't sustain additional animals. That's because urban fox populations are regulated by the same factors that regulate populations of other wild animals: food availability, disease outbreaks and competition for territories. Bristol's high fox population, for example, was decimated by an outbreak of sarcoptic mange in 1994, which caused it to crash by more than 95 per cent. Ten years later, Bristol's fox population density was only 15 per cent of what it had been immediately prior to the outbreak, although it is now increasing back to pre-disease levels.

Like the cliff swallows, urban foxes have adapted to living alongside us in many ways, resulting in some differences with their rural counterparts. Urban foxes have smaller, more fluid territories with larger social groups, they are more promiscuous and a larger proportion of their diet is made up of scavenged food. Another adaptation, Scott points out, is that urban foxes make full use of the three-dimensional environment – all foxes are good climbers, but in the wild they rarely climb trees unless they need to. In cities foxes have adapted, incorporating roofs and other elevated spaces into their territories. It makes sense: this otherwise unexploited habitat is usually safe and sheltered so can be a great place to lay up during the day. Vixens have even been known to have litters up on roofs, whereas they'd never have them up a tree in the wild.

Surviving here likely requires certain cognitive traits, such as rapid learning ability, good spatial memory, innovative foraging and flexibility. From Scott's perspective, urban foxes have met the challenge: 'They have managed to learn how we function and how we behave; they've adapted their behaviour around us to exploit what's there and they've done it quite successfully.' We know that foxes can rapidly learn about certain aspects of human life. Traffic is one, although in many cases it's a one-shot learning trial. Outside of the breeding season, foxes tend to avoid crossing roads and they manage this by forming territories made up of several contiguous suburban gardens. Baker and the Bristol team found that if foxes do cross roads, they're more likely to do it after midnight compared with earlier, which coincides with the lowest traffic volume. Since foxes that were not road-savvy wouldn't have survived in the city for long, traffic mortality may have exerted a selective pressure on urban foxes for 'road awareness'.

Cities are complex environments, full of new opportunities, but also rapid and unpredictable changes. To thrive here, we might expect evolution to favour animals that can solve problems quickly, something that seems consistent with our tales of the fox. And yet, scientific studies of fox cognition are few and far between. Scott and her students have run a pilot study, putting out feeding boxes and asking whether the foxes will pull a string to drag in food when it is out of reach, as has been found for dogs. The few animals that they've tested have worked it out, with some of them 'sort of' solving the problem when the strings are crossed (this asks whether the animal understands that the connectivity between the string and treat is important). 'I think they have the capacity to be very intelligent,' Scott says. 'They're good at problem-solving, probably at least as good as dogs. But within a fox population they're all very different.' She and her students tried to compare the problem-solving skills of urban and rural foxes, but they ran into a sizeable stumbling block. Rural foxes are too fearful to even approach the testing apparatus, which harks back to Macdonald's failed attempts to trap the animals. They need to find a population of rural foxes that aren't already habituated to people, but that aren't so fearful that they won't engage with the apparatus. And so far, although the team might be able to collect data on many urban animals, they're lacking this comparator group.

Scott's experience suggests that urban foxes may differ from rural foxes in personality traits, including neophobia, but it's the same story regarding sample size – rural foxes are too fearful of strange objects in their territory. One of the clues ties in with another well-known behaviour of foxes:

playfulness. Again, this is not something restricted to urban animals: foxes are naturally playful from the earliest age, like many social mammals that exhibit play-fighting to learn skills and establish dominance. However, urban environments offer something else to a curious fox, in the form of myriad novel objects that can be found, primarily in our gardens. In 2020, a fox in Berlin was reported to have stashed close to 100 flip-flops that he had stolen from people's gardens, and I have certainly experienced the confusion of finding my trainers, left on the back doorstep after an intense training session, at the end of the garden the next morning. Foxes have likely been using manmade objects as playthings for thousands of years and it's a lovely thought that in Aesop's time, foxes may have made the same nuisance of themselves by stealing sweaty sandals as they do today. Since then, of course, foxes have been persecuted so intensely that rural animals are deeply fearful of humans – they carry the genes of the animals that survived, after all. But in just 100 years of living alongside people, is it possible that urban foxes are losing their fear?

Scott and team will keep trying to collect data from rural foxes, but in the meantime they're taking a different approach to evaluate personality traits. Working with welfare organisations, they have recorded how different aspects of personality developed in a litter of motherless cubs that was brought into a shelter. In collaboration with a television company, the team followed the litter's development and documented their personality traits. They have found striking individual differences in fear, neophobia and problem-solving. What's more, the traits seem to stay consistent as the cubs grow up – bolder individuals tended to stay bold, explorers tended to remain exploratory and so forth. How the development of these personality traits varies in rural and urban foxes is another question; much more research is needed on the development and cognition of foxes.

On the physical side of things, there's a common misconception that urban foxes are smaller, scrawnier animals than their plump, glossy rural counterparts. This is a myth and until recently, fox experts would tell you that there is no physical difference between urban and rural animals; but a recently published study suggests this might not be the case after all. The study analysed anatomical features of skulls collected by Harris in London in the early 1970s and revealed some differences. Urban foxes had shorter, wider skulls compared with foxes living in the surrounding, rural boroughs; they also had a larger nasal region and, contrary to predictions, a smaller braincase, particularly in males. The latter point is particularly striking because it contrasts with previous research of museum skulls, which showed that urban populations of certain small mammal species had larger brains compared with rural populations. Bigger brains may enable greater behavioural flexibility, which seems to make sense for urban life, so what's going on with the foxes?

One possibility is that a shorter, wider snout is simply a by-product of dietary changes promoting greater muscles in the jaw, which may be advantageous for processing human leftovers and litter. And yet, much of the food that foxes scavenge in towns requires no additional force to bite through than in their rural diet. A more compelling hypothesis is that the skull changes may reflect the first signs of domestication syndrome in red foxes. If animals living in human environments show greater boldness and tolerance of humans, as they are expected to in order to benefit from human resources, this may promote physical changes just as Belyaev's silver foxes did when selected for tameness. Skull changes are one of the traits that make up domestication syndrome, as is a smaller braincase; some researchers have suggested that a shorter, wider skull allowed dogs to produce a greater number of facial expressions and therefore to communicate with humans. It's an intriguing finding and only time will tell whether urban foxes have started on the path to self-domestication.

★ ★ ★

Just like its bigger cousin, the wolf, the fox is a creature that polarises opinion. Tolerance has increased in many places, but foxes are still frequently regarded as vermin. In Iceland, for example, arctic foxes have been persecuted since people first settled. The arctic fox is a smaller, fluffier animal than the red fox, adapted to survive prolonged sub-zero temperatures by the best insulated fur coat of any mammal.[*] In 1295, a law was passed that fined farmers who did *not* kill at least one adult arctic fox every year, and through the Middle Ages fox pelts were used as legal tender. It is only since 1994 that the foxes have received some degree of protection; today, extermination is no longer the goal and despite steady hunting pressure, the fox population is stable.

Red foxes have also been hunted for centuries in one form or another, although it is widely agreed that the first record of foxes being hunted by hounds was in 1534, when an English farmer wanted rid of the animals. It resurfaced in the seventeenth century and by the early nineteenth century, following intense persecution by gamekeepers and farmers, wild fox numbers were so depleted that foxes had to be imported from continental Europe for people to hunt. They were distributed from Leadenhall Market in London, where a 'Leadenhaller' sold them alive in bags and the fox was only released from the bag at the time of the hunt. It is said that demand became so high that gangs would even catch foxes from one part of England and sell to hunts in other areas, and

---

[*] This fur has put intense hunting pressure on Arctic foxes throughout their range. In North America between 1919 and 1984, for example, approximately 40,000–85,000 animals were killed each year.

some landowners were forced to pay 'protection money' to prevent gangs targeting the foxes on their lands.

It was hunting that brought the fox to Australia's shores – a particularly ill-thought through decision in the mid-1800s. By 1870 wild populations were well established and within 100 years foxes had spread out across most of Australia, including several cities. The consequences have been disastrous for native ground-nesting birds and reptiles, which have no evolved defences against this adaptable, tenacious predator. Even worse, the only natural predator of foxes, the dingo, has itself been the subject of intense persecution, allowing the fox population to expand almost unchecked.

In the UK, the red fox and European badger are our biggest terrestrial carnivores. An adult fox is typically about one-fifth the weight of a Labrador and the badger's a bit bigger. Some countries have lions, tigers, crocodiles or bears roaming the landscape; all animals that could quite easily kill a human and all that are worth fearing. Why then, are so many people, who have no direct experience with foxes, so uneasy about encountering one? One possibility is that urban foxes have a historical reputation for being a nuisance and were often, erroneously, blamed for raiding bins. In fact, studies of the Oxford and Bristol foxes showed that cats and dogs were often the culprits too. Another concern is disease transmission: as we heard already, sarcoptic mange can run rampant through fox populations and this can also be transmitted to dogs, as can fox tapeworm. Red foxes are a reservoir of rabies, and while much of Western and Central Europe has achieved rabies-free status following introduction of a highly effective oral rabies vaccine, there remain areas where disease transmission is a real concern.[*]

---

[*]Rabies fears are unfounded in the UK: we have been rabies-free since the early twentieth century, apart from a form carried by bats called bat lyssavirus, which is very low risk unless you are regularly handling wild bats.

Another, more troubling, possibility is that foxes are wild animals and we are losing our connection with the wild at an alarming rate. For people who don't have the opportunity to regularly engage with nature or wildlife, a lack of knowledge or understanding might promote feelings of fear when confronted with a fox. It's a wild carnivore, so theoretically it could do some damage. But for the vast majority of cases, the only time a fox will get aggressive is if it's cornered and feels threatened enough to take the risk. This small, graceful animal wants to bump into you about as much as you want to bump into it.

For Scott, a lot of it comes from the legacy of fables and other stories that have characteristically portrayed the fox in the same way. 'You have this established villainisation, which I think has come from hundreds and hundreds of years of stories, and it is such a shame. It's really hard to try and persuade people otherwise.' She tells me of one example where a person told her that a fox had killed a cat, but when asked for details not only did the person not see the incident, but the cat was found next to a road. Scott suggested the cat may have been hit by a car, but the person was adamant that because they'd seen a fox in the street the previous night it must be the culprit. It's a staggering leap, another example where our brain jumps to conclusions to confirm an existing belief, rather than rationally considering the evidence. In fact, urban foxes and cats usually ignore each other and if there are aggressive encounters the cat usually wins. Foxes can injure and kill cats, but it's not normal behaviour; cats are apparently strange and confusing adversaries because they don't respond like wild predators.

Certain sections of the media fan the flames by using consistently negative language and imagery in their stories about foxes. Scott has one newspaper example she likes to discuss with people in which, within the first three sentences, eight negative terms are used. This particular 'news' story was about a group of people who are said to have undergone a 'four-hour long ordeal' inside a pub because a 'vicious' fox was 'terrorising' them in the car park. For Scott, nothing about it makes sense: 'Why is this newsworthy and why use eight negative connotations about an animal that was in a car park?'

Perceptions have changed in recent years; more people are sharing positive stories of the foxes in their garden and in a recent study of public perceptions, nearly 90 per cent of respondents said that seeing a fox enriched their lives. Favourable views are presented in documentaries, on social media and in books, meaning that liking foxes has become more socially acceptable. The trouble is that the dwindling minority of people with anti-fox views have louder voices and often get a platform from which to shout them, as with so many areas of life.

Fact or fiction?

The fox. Sharp, quick-witted, crafty. Persistently typecast as a cunning villain, the fox is the animal that talks its way out of mischief while talking others in. Aesop's fables are consistent in their portrayal, with 'The Fox and the Crow' perfectly summing up the fox's cunning character – but I could have chosen to include any of the others here. It seems that even at this point, some 2,500 years ago, foxes had a reputation. We can't ever know how far back that belief may stretch, but what is very clear is how strongly it's endured.

Scientific studies of fox behaviour have mainly been tied in with its status as a predator, a disease vector or an invasive pest. We know a lot about what foxes eat, what diseases they transmit and how their populations work. Dedicated studies into fox cognition, on the other hand, are few and far between. Scott's research suggests that foxes may have similar cognitive abilities to dogs when it comes to solving a string-pulling problem, but so much is unknown about the mechanisms underpinning fox intelligence that it's impossible to make firm conclusions about the reality of Aesop's fox. Certainly, the red fox is an adaptable, opportunistic animal; it is as at home hunting lemmings in the Arctic Circle as it is catching moths in rural pasture and scavenging in people's gardens. To people who say that foxes belong in the countryside, consider that to be just one of the fox's many habitats; it is no more or less its 'rightful' place as is the forest, desert or tundra. Towns and cities are simply another habitat to add to that list, meaning that foxes 'belong' here just as much as they 'belong' in any of the other places.

We don't experience a city like a fox does: its heightened senses take in sights, sounds, smells, opportunities and dangers. Surviving in this environment requires certain skills. And yet, foxes aren't just surviving here, they're thriving. And that has taken something more. They have learned how we function, and they've used this information to adapt to living in the city. They've figured out when and where to roam to avoid predictable danger and to get an easy meal. They've adapted their social behaviour, become more tolerant of bigger groups and neighbours, because all can benefit from the abundant food we provide. We may see a concrete jungle, but foxes see a habitat. It's a fantastic case-study of how humans and wildlife can coexist.

The fox traits of adaptability, opportunism and flexibility enabled its spread around the world. Urban colonisation was just one more chapter in fox evolution. But working out how to live alongside us and exploit our habitat isn't the same as being a crafty rogue. So how should we think about foxes? Over to Scott: 'Part of the solution is to try and understand a fox as an animal, rather than anthropomorphising it into a villain. To appreciate it for what it's trying to do and understand that it's actually a really good survivalist. You have this amazing opportunity to have this carnivore living right next to you and I think that's a gift.'

CHAPTER SIX

The Lion and the Shepherd

A Lion, roaming through a forest, trod upon a thorn. Soon afterward he came up to a Shepherd and fawned upon him, wagging his tail as if to say, 'I am a suppliant, and seek your aid.' The Shepherd boldly examined the beast, discovered the thorn, and placing his paw upon his lap, pulled it out; thus relieved of his pain, the Lion returned into the forest. Some time after, the Shepherd, being imprisoned on a false accusation, was condemned 'to be cast to the Lions' as the punishment for his imputed crime. But when the Lion was released from his cage, he recognized the Shepherd as the man who healed him, and instead of attacking him, approached and placed his foot upon his lap. The King, as soon as he heard the tale, ordered the Lion to be set free

again in the forest, and the Shepherd to be pardoned and
restored to his friends.

'Wait here,' murmurs our safari guide, 'Top Gun', 'I need to
check.' He disappears into the bush, and we exchange nervous
smiles. It's a couple of hours after sunrise in Botswana's
Okavango Delta and we are on a walking safari. After
explaining the explosion of hippo dung spattered outside our
camp,[*] Top Gun spotted some lion prints so, naturally, we're
following them. The son of a Bushman, he has lived his
whole life on the edge of the Delta and he knows his stuff. So,
when he tells us that the prints are no more than an hour or
two old and that they come from a 'really big group', a little
pit of anxiety ignites in my gut. This part of the plain was
recently scorched in a wildfire, so the prints – which are
worryingly large – stand out clearly in the soft ash. We learn
that cats' paw prints differ from dogs'; cats have three lobes in
their rear pad, whereas dogs only have two.

  We've stopped near some spiny scrub, the savannah
stretching off to our left. The grass is long enough that lions
could be mere metres away and we'd have no idea. In
hindsight, not the right thing to say. Top Gun has emerged a
short way away; crouched down, he's absolutely focused on
the ground. We couldn't be in safer hands, but we're all still
terrified about the prospect of meeting a large pride of lions
in the middle of wild Africa. Yesterday, on the far side of this
plain, we were chased by a protective hippo mother as we
watched a pod wallow in a lake. We ran, full pelt, over the

---

[*] The previous night, we listened to the groaning, growling roars of
hippos fighting in the waterway just outside our camp. This dung will
have been produced by the winner of that contest, staking a claim to his
territory. Hippos have a niche skill for spraying dung, whirling their
stubby little tails as a kind of fan that ensures the dung gets sprayed
around for maximum effect.

plain towards where we are standing now. It is sobering to think that we could have been running from the jaws of one dangerous beast and straight into the jaws of another.[*]

Top Gun is back and he's looking relaxed. 'This way,' he points. 'Banana banana.' This is his way of saying that everything is good, so we smile, relax and continue our walk. Still, no one's hanging around at the back.

The African lion is one of the Big Five, a term originally coined by game-hunters for the five most difficult animals to hunt on foot in Africa. Today, of course, it's ubiquitous with safari holidays and millions flock to the continent, using cameras to try and 'shoot' the Big Five for themselves. The other four are the elephant, rhinoceros, leopard and buffalo, all of which are notoriously dangerous animals. Elephants, rhinoceroses and buffalo are huge, powerful juggernauts, and you'd be lucky to survive being charged by one. But they're not predators, so they'll usually attack only when they feel threatened. Leopards, like lions, are powerful apex predators, and they do attack and kill people. But for me, the lion is scariest of all.

As a muscular, powerful, ferocious hypercarnivore,[†] a single lion could take down an adult human with ease. But that's nothing compared with the teamwork of a pride, which can tackle fully grown buffalo, giraffes and even young elephants. This is where the fear factor comes for me and it's well founded – lions are one of the deadliest animals on the planet. In 1898, two lions in the Kenyan region of Tsavo terrorised railway workers over a period of nine months before they were shot. Named 'The Ghost' and 'The

---

[*] One might say 'out of the frying pan and into the fire', an expression we also owe to Aesop, for his fable of 'The Stag and the Lion'. In this, a stag that was being pursued by hungry dogs hides in a cave that, you guessed it, was also a lion's lair.

[†] Meaning that more than 70 per cent of their diet is meat.

Darkness', these large, maneless males are thought to have killed and consumed as many as 135 people. Humans are easy – but not normally preferred – prey for lions. It's thought that the Tsavo lions had tooth and jaw problems, which made chewing through the tough hides of their usual prey excruciatingly painful. Lion attacks aren't a thing of the past, either. Between 1990 and 2005, there were close to 900 lion attacks on Tanzanian villagers, 563 of which were fatal. And in July 2018, a group of rhino poachers broke into the Sibuya Game Reserve in South Africa's Eastern Cape, armed with a high-powered rifle and an axe. The remains of two, possibly three, of these men were subsequently discovered by a guide; the reserve's resident pride had found them first.

Due to its strength and majestic appearance, the lion became king of the beasts a long time ago and today it remains one of the most iconic animals on the planet. Few other animals have had such an impact on the myths and legends of the people it encountered. From the terrifying and apparently indestructible Nemean lion, slain by Heracles in Greek mythology, to the ancient Egyptian lioness-headed goddess of war, Sekhmet, who protected the pharaohs and was responsible for the annual flooding of the Nile. The Nubians worshipped a lion-headed god of war called Apedemak, thought to be the origin of the Egyptian equivalent, Maahes. In Mesopotamia, lions were important symbols of royalty and Assyrian kings bred them in captivity as early as 850 BC. Lions were also the ultimate opponent against which man could prove his strength and courage: for this reason, the Romans slaughtered many thousands of the animals in their arenas.

Lions have always signified great strength, bravery, courage and nobility; everything about them adds to this portrayal, from their place at the top of the food chain, to their 'crown' of fur, the mane, to the perfectly unhurried, composed way that they go about their business. The first recorded coat of arms, granted to Geoffrey Plantagenet in 1127, featured four red lions, subsequently evolving to the three lions first displayed by none other than Richard Lionheart, King Richard I. The term 'pride', first recorded in the fifteenth-century *Book of St Albans*,* seems very apt and there is no evidence that it arose from anything other than an anthropomorphic view of what it is to be a group of lions. As written by Alfred Pease in his *The Book of the Lion* (1913):

> If you would flatter a king, you say he is 'lion-hearted'; if a man, that he is as 'brave as a lion'. If a nation desires to impress the world at large without any false modesty it assumes this great cat as its badge, and when so many nations have taken the lion as their pet symbol that the thing begins to get monotonous, the rest fall back on the eagle or bear.

Townsend's translation of Aesop's fables contains 35 fables with 'lion' in the title. Their characters are broadly consistent: powerful, noble animals that can take what they want without fear of retribution. Aesop's lion doesn't have the same crafty wit as the fox, because it relies on its physicality and status as the dominant beast. And yet, the lion is not always a tyrannical leader; this fable and 'The Lion and the Mouse' depict reciprocal helping towards another individual, clear cases of 'you scratch my back and I'll scratch yours'. What's less clear is whether this kind of reciprocal helping happens in nature: in lions or any other animal.

★ ★ ★

---

*Along with other delights, including a shrewdness of apes, exaltation of larks and chattering of choughs.

Lions belong to the cat family, which is scientifically called the *Felidae* and, according to the most recently updated taxonomy in 2017, includes 41 species. Seven of these are the 'big cats': the four belonging to the genus *Panthera* (lions, tigers, leopards and jaguars) plus snow leopards, clouded leopards and cheetahs. All the others are classed as 'small cats', although confusingly the distinction isn't based on size but on differences in their vocal anatomy, which make small cats unable to roar. One notable size exception to the big cat grouping is the mountain lion (aka cougar or puma), which is a larger animal than both the cheetah and leopard.

Felids have an extensive fossil history, with the earliest thought to have appeared in the region of 35–28.5 million years ago, i.e. long before the appearance of hominids. Their closest living relatives are the Asiatic linsangs: slender, tree-dwelling carnivores that look a bit like elongated cats with stretched-out snouts. Many other species of cat have walked the Earth and are now extinct, including a group of sabre-toothed cats in the genus *Smilodon*, which had 7-inch-long canines and a gape nearly twice the size of modern cats. The three *Smilodon* species went extinct around 13,000 years ago. There used to be other lion species too. In fact, lions were once the most widespread mammal, covering Africa, northern Eurasia and Central America. The European cave lion, for example, ranged across Europe and Asia into Alaska, and is represented in stunning form in the Chauvet caves. Here, lions are depicted hunting and consorting, startlingly similar to the behaviours of modern lions. Whoever painted them must have been as inspired by the animals as they were knowledgeable; they would have lived alongside these creatures and competed with them for food. The European cave lion disappeared at the end of the last ice age, some 14,000 years ago, while the last evidence for North American cave lions is about 11,000 years ago.

Some extinctions have been more recent. The Barbary lion was a distinct geographic subspecies of African lion, confined to North Africa. The largest subspecies of lion, it was a magnificent, dark-maned creature (likely the inspiration for lion symbolism and motifs), and almost certainly the lion exploited by Romans in gladiatorial events and later menageries of medieval Europe. The Romans killed many thousands of Barbary lions for their 'games', but it was the arrival of European colonial hunters that finished them off, driving them to extinction in the wild in the mid-twentieth century.* There were also lions in Greece: the historian Herodotus considered the animals to be common around 425 BC, while by Aristotle's time, around 300 BC, they were considered rare and by 100 AD they were gone. In fact, in the past 150 years, we have seen extinction in the wild of the Barbary lion in North Africa, the Cape lion in South Africa and lion populations in the Middle East – the last was shot by a hunter in Iraq in 1918.

In contrast with canids, where foxes are the odd ones for not living in social groups, among felids going solo is the norm. This is true of the tiniest 1kg rusty-spotted cat, right up to the 300kg Amur tiger, the largest cat in the world. Even your pet cat, although it may happily live with you and tolerate any other cats that you own, is genetically almost identical to the African wildcat, which is a solitary hunter.

Lions are the oddballs within the cat family because they live in permanent social groups, called 'prides'. Numbering between two and 21 individuals, they comprise a permanent core of lionesses, their dependent offspring and a ruling coalition of males. Lionesses are the heart of the pride and a

---

*Captive individuals remained in the private collections of regional nobility.

wonderfully egalitarian group: they hunt together, share food and all of them breed, in contrast with other social carnivores, such as hyaenas, in which a dominant female monopolises breeding. About three-quarters of female cubs will stay on in their mother's pride once they reach adulthood; the others will strike out to form new prides. Male cubs, on the other hand, don't stay in the pride and neither are they recruited by their dads into the ruling coalition; instead, at the age of two to three years, all young males are ousted and commence a roaming, nomadic life. If they have brothers or cousins, they leave together and remain in a related gang (literal 'brothers in arms'); if not, they wander the plains until they find another male or males to buddy up with. For nomadic males, their ultimate life goal is taking over a pride (or prides), because that's their only route to fatherhood; their lives revolve around either taking or holding on to control of a pride.

Lionesses have primary responsibility for hunting and rearing cubs. Lithe, patient ambush hunters, they are impeccably camouflaged when they creep through the arid vegetation – and they're fast. Not 'cheetah fast', but once they've been spotted by their intended prey, they can pump the gas and hit 65km/h over short distances. Males, on the other hand, are considerably bigger, heavier animals and they're not really built for hunting. Described as 'partial parasites', they spend most of the day lazing around, rousing themselves when the lionesses have made a kill and bullying their way to the front of the queue. In fairness, their size and bulk make them slower and less agile than the females, so they'd find it much harder to creep along the ground to stalk prey. And in any case, that mane's a bit of a giveaway. George Schaller, who was the first to document the ecology and behaviour of lions in the Serengeti, summed it up nicely when he wrote that a male trying to hunt in the day looks like a 'moving haystack'. That doesn't mean he can't or won't hunt – his size and strength allow the pride to bring down large, dangerous prey like adult buffalo, a hugely important food source during the breeding season.

Males aren't top hunters because they're built for something else – fighting. The pride's protectors, they need to be big, strong and ready to respond to challenge at any time. Those roaming, nomadic coalitions will skulk on the periphery of territories, biding their time until they make a challenge – and the stakes couldn't be higher. Lions, while being strongly bonded to the individuals within their pride, have zero tolerance for any other lion and fighting is the biggest cause of natural mortality. Lionesses will take out 'the competition' by attacking and killing lionesses from neighbouring prides, and male takeovers are brutal, often resulting in blinding or death for the losers. It doesn't stop there either: when new males take over a pride, they systematically kill any cubs young enough to still require nursing and evict any sub–adults, which is often a death sentence in itself. The data show that one in four of all cub deaths results from infanticide. It's therefore in the lionesses' interests to ensure their males are well fed, well rested and in optimal condition to respond to intruders: not just other lions, but hyenas too. They rely on the males to keep their cubs safe.

Infanticide is common in lions, just as it is for many species of mammal, including dolphins, baboons and rodents. Even our closest living relatives, chimpanzees, do it. It can be tempting to demonise males for this act, because we can't help but think about it through the lens of our own behaviour. But lions are not humans and understanding infanticide means understanding the intricacies of lion biology. It all boils down to sex. Or rather, the lack of it. Because while a lioness is nursing her cubs, she has no interest. Her priority is protecting and rearing her young until they reach independence, a period of at least 18 months, and during this time she will aggressively fight off amorous suitors. For new pride-holders, this is a big problem. They may only get one stint of tenure in a pride, one opportunity to pass on their genes, and they don't know how long it will last. Studies have found that coalitions of two to three males have a 50 per cent chance of holding a pride for two years, while for coalitions of four to five males this might be more like 80 per cent. New pride-holders therefore face a very real pressure to start breeding as soon as possible: they can't afford to stepfather unrelated cubs to independence because they might be overthrown by then. Killing the cubs brings the females back into oestrus in a matter of days, making it a bit of an evolutionary no-brainer.

When a female shows signs of becoming reproductively active, the first male on the scene forms what is known as a 'consortship' with her. A consorting lion has been likened to a bull elephant in musth: he has just one thing on his mind and he won't tolerate anything getting in his way. The lioness is in oestrus for four to five days and the male needs to keep rivals away from her for the entire duration, something he achieves by sticking by her side and mating with her every 25 minutes or so. This is another reason why a male needs to conserve his energy! You may have seen from documentaries that when a male dismounts, the lioness often snarls and swats at him. That's because lions, like all felids, have backward-facing barbs on the penis (called 'penile spines'), which rake the female's vagina as it moves.[*] It's an intense period and the animals often don't eat, leaving them weakened by the end.

Males can come into conflict over mating. Coalition partners who have fought shoulder-to-shoulder to keep

---

[*] The testosterone-dependent spines are made of keratin and it's thought that the 'raking' may serve to stimulate the female's reproductive tract and trigger ovulation.

control of a pride can turn on each other, and vicious fights can result in wounds to the face and eyes. Conflict is reduced by the fact that lionesses often come into oestrus at the same time, which, particularly if the pride is large, means there's no reason to fight for sex. What's more, females will often mate with more than one male, which also reduces any conflict towards her cubs, since the males don't know which they have fathered. When conflict does occur between male partners, they may reconcile by nuzzling each other and, in some cases, mounting. It looks sexual, but is more akin to having a good man-to-man hug to smooth over tensions.

★ ★ ★

Clearly, there is a lot going on in a pride. Dramas are being played out across the savannah; family feuds and neighbourly disputes abound, just like any good soap opera. And the question of why lions show all this social activity, when all other cats stayed true to their solitary roots, is one that has fascinated researchers for decades. The Serengeti Lion Project has provided many of the answers.

The study was initiated in 1966 and continues to this day, making it one of the longest-running population studies of any carnivore. It was the vision of John Owen, then director of the Tanzania National Parks, who invited Schaller and his family to the newly established Serengeti Research Institute to study the park's lions. Owen tasked Schaller with understanding how lion predation influenced the dynamics of the huge herds of wildebeest, zebra and gazelles that concentrated in the park. Schaller, who has referred to himself as a 'feral biologist' and is one of the world's most influential wildlife conservationists, spent three years studying Serengeti lions, notching up a mighty 149,000km of travel and almost 3,000 observation hours. Schaller spent so long watching many of the same animals, he was able to identify some 60 lions by their natural scars and markings alone; for about 150 others, he relied on ear tags. His observations were published in *The Serengeti Lion* (1972) and provided a solid foundational layer into lion ecology and behaviour, paving the way for the subsequent decades of research that continue to this day.

The project was inherited by Brian Bertram in 1969, then Jeanette Hanby and David Bygott in 1974, and Craig Packer and Anne Pusey in 1978. Packer, now a distinguished McKnight professor of ecology, evolution and behaviour at the University of Minnesota, USA, has been working with lions ever since. Pusey, who is the James B. Duke professor emerita of evolutionary anthropology at Duke University in North Carolina, USA, and director of the Jane Goodall Institute Research Center, has remained focused on the evolution of social behaviour, but subsequently working with primates. For the past 25 years, she has worked almost exclusively on the Gombe long-term chimpanzee project.

Packer's journey to Africa began in 1972 when, as a pre-med student, he took the opportunity to spend a year at Gombe. At that time, he'd barely heard of Jane Goodall and her pioneering chimpanzee research, so he chose to work on a less popular project on olive baboons. At Gombe, he met Pusey, who had joined Goodall as a research assistant in 1970. Packer subsequently ditched medical school and returned to conduct his PhD research with the Gombe baboons in 1974, while Pusey conducted an ethological project on adolescent chimpanzees. A few years later, the opportunity came up for Packer and Pusey to join the Serengeti Lion Project. Packer has described being initially underwhelmed by the animals, finding their habit of sleeping for much of the day very boring compared with the baboons. Fortunately, the pair recognised that with 12 years of demographic data behind them, this was a fantastic system to study the intricacies of social behaviour. In 1986, at the

University of Minnesota, Packer set up the first research centre devoted to the study of lions. More than 40 years later, he's still leading the research, directly in Kenya and other countries, and indirectly through students and collaborators in the Serengeti and Ngorongoro Crater.[*]

Individual lions can be identified by their unique whisker spot patterns and, as they grow older, the scars and notches that accumulate from fights. This method is fine for watching lions during the day but, as Packer and Pusey learned, a lot of interesting behaviour happens at night. And as stealth hunters, lions don't want to be seen. Schaller, amazingly, was capable of identifying individual lions on moonlit nights, but that was challenging enough even for someone with his skills. Fortunately, reliable radio telemetry was introduced to the study in 1984 and productivity skyrocketed. Since then, one female in every pride is fitted with a radio collar, allowing the team to collect routine data on pride locations and to get an idea of where they're hunting overnight. It's an invaluable tool when pride territories can be as large as $400km^2$.

I was delighted to spot one of these females a few years ago in the Serengeti, while watching a pride lounging alongside a river. Early that morning, we had been startled and a little bit terrified[†] to hear the deep, guttural, grunting roars of lions, clearly just outside our unfenced camp. In fact, we were amazed to learn that under some conditions, lion roars can be heard 6–8km away (and lions, with superior hearing, might

---

[*] The Tanzanian government banned Packer from returning to the country in 2014, the result of his outspoken remarks about corruption in the government-run trophy-hunting industry.

[†] The previous night our camp had been surrounded by a hyaena clan. There is something about shining a torch out into the black and seeing eyes glowing back at you, while hearing that laughing whoop, that chills right to the bone. It's fair to say that a few of us were a little jittery the next morning.

hear them up to 12–15km away) and that these were at least a couple of kilometres away. As we drove away a little later, we happened on what must have been the same pride and were granted brief access into their world; a privilege that absorbed my every attention. Lions have what is called a 'fission-fusion' social structure, which means that although they all belong to the same pride, they don't spend all of their time together. Pairs or trios may split off to hunt or nurse young, but they come back together and when they do, they reinforce their social bonds with greeting displays. We watched as a male approached two returning lionesses and nuzzled their heads and necks, a behaviour so familiar to any cat owner who experiences the same every morning.* A sub-adult male was pointed out, our guide chuckling at his short scruffy mane. And then a lioness came into sight, a thick grey band around her neck – part of the Serengeti Lion Project!

When Packer and Pusey took over in 1978, they had a straightforward aim: to work out why lions had bucked the trend of all other cats and teamed up to hunt, rear young and defend each other. 'All this togetherness did not make much evolutionary sense,' the pair reflected in a later review. At the time, the evolution of social behaviour was becoming a major focus for biologists and at the heart of this was the recognition that individuals could interact with others in two broad ways: by cooperating or competing. Remember that the Darwin–Wallace theory of evolution by natural selection proposed that traits which increase an organism's likelihood of surviving and reproducing will be favoured. From this, it seems that the world should be dominated by selfish, deceptive behaviour, because every individual is competing with every other individual to survive. In fact, there are many behaviours in the natural world that cannot be explained by competition

---

*The Wimpenny word for this is 'smooging'.

and scientists have long puzzled over how to apply evolutionary theory to cooperative behaviour. Particularly problematic in the early days were social insect colonies, such as ants, termites or honeybees, in which one or a few breeding queens are surrounded by thousands of sterile workers, foregoing their own reproduction and even sacrificing themselves to defend their queen and colony. This apparent paradox of evolutionary biology presented complications for Darwin, being 'by far the gravest difficulty, which I have encountered'. Darwin's solution was to consider that selection may also act at the family level, so by foregoing reproduction an individual ant helped the group by preserving adaptive traits of different castes, leading to continued selection of the 'stock'.

In the early twentieth century, cooperative behaviours were assumed to be evolutionarily beneficial because they were for 'the good of the group'. But by the 1950s there were many who rejected the theory of group selection and refocused attention to the level of the individual and the gene. This culminated in paradigm-shifting new theories and the beginnings of a new approach to the study of animal behaviour – one that concentrated on the study of evolutionary adaptations and the function of behaviours for maximising individual fitness. Behavioural ecology focused on working out the costs and benefits of behaviours to understand how they enhanced an individual's survival and likelihood of breeding. Cooperative behaviours were still a problem, particularly the form of cooperation known as altruism, where one individual experienced a fitness cost for helping another by providing them with benefits. How could such behaviour evolve when individuals should be adapted to maximise their own fitness?

Prominent mathematicians Ronald Fisher and J. B. S. Haldane had explored the basis of cooperation in the 1930s and they hit upon a crucial point – that when thinking about the genetic evolution of any trait you cannot just look at how it impacts individual reproductive success. That's because we share copies of our genes with our relatives: we inherit 50 per cent of our mother's genetic material and 50 per cent of our father's, with 50 per cent shared with each of our full siblings. To properly evaluate the evolution of a trait, you therefore need to know how it impacts the fitness of relatives too, because even if it doesn't directly benefit you, it might enhance their survival – and since they share copies of your genes this leads to indirect genetic benefits. Haldane is widely credited for popularising this concept, famously joking to his mentee John Maynard Smith that he would lay down his life 'for eight cousins or two brothers'.[*]

Maynard Smith subsequently defined the term 'kin selection' in 1964 and that same year theoretical biologist Bill Hamilton published the first mathematical model to address the problem of altruism. Known today simply as 'Hamilton's rule', it defined the field. Altruism, Hamilton explained, will only be favoured if the benefit (B) to the altruistic individual outweighs the cost (C) of helping. Most of the time this condition won't be met, because time spent helping another individual is time spent away from beneficial activities like feeding, mating or watching for predators. But, if you plug in the relatedness (r) of those individuals, things might change. By working out the fitness benefits and costs of helping at a direct and indirect level, you get what is called 'inclusive fitness'. Hamilton's rule predicts that altruism is favoured if $r \times B > C$: it means that even costly behaviours can evolve by natural selection if the beneficiary is a close relative, because overall inclusive fitness is still high.

---

[*]That's because a sibling shares 50 per cent of your genetic material while a cousin shares only 12.5 per cent.

Inclusive fitness was thought to have solved the paradox of cooperation between relatives, including the ultra-cooperative social insects (termed 'eusocial'), because of their peculiar haplodiploid sex determination. Males (drones) develop from unfertilised eggs and possess only one set of chromosomes, which is termed 'haploid'. Females (workers) develop from fertilised eggs and possess both sets of chromosomes, meaning they are 'diploid'. The unusual inheritance that results means that females end up being more related to their sisters than they are to their offspring, leading to evolution of a caste structure, where it's in the females' genetic interests to help their mother produce more females (i.e. sisters). Eusocial insects have taken cooperation to extremes, pursuing an apparently bizarre evolutionary path, but one that is certainly not the norm.

It's not as simple as that, of course – there are still many questions around the evolution and maintenance of cooperation, and the really hard one concerns why unrelated individuals are willing to accept the cost of helping. It was a puzzle that left biologists scratching their heads until 1971, when Robert Trivers published his pivotal theory of reciprocal altruism (today, more commonly referred to as 'reciprocity'). Trivers suggested that altruistic behaviour could be favoured among unrelated individuals if the 'helper' was paid back at a later date and outlined three key conditions under which reciprocity could evolve. First, the cost to the helper of helping must be low; second, the help must confer high benefits to the receiver; and third, there must be sufficient opportunity for both individuals to repeatedly interact. Trivers' theory was a huge step forward, although many questions remain. Nonetheless, it set out several testable predictions and, armed with theories and predictive models, behavioural ecologists now needed the data to test them. Working out the costs and benefits of cooperation in animals became a dominant research focus in the 1970s and numerous studies were set up. Many of these continue to the present, providing decades of high-quality data on the behaviour and ecology of animals.

Back to the Serengeti, where Packer and Pusey were puzzling over the evolution of lion cooperation. The most popular hypothesis, and the one that had been favoured by Schaller, was that prides had evolved so that lions could hunt bigger prey, improving their chances of survival on Africa's plains. Lions do hunt together and the prey that they capture can be considerably larger than one animal, so intuitively this makes sense; but the data did not fit. For one, you would expect that larger prides should feed on larger prey than smaller prides and this isn't the case. You would also predict that a solitary lioness would be limited to catching warthogs or other small prey. Not true either – a single lioness is perfectly capable of bringing down a wildebeest or impala, and during a group hunt it's often one lioness that does most of the work. While lions can and will cooperatively hunt, they don't do it enough for this to explain why prides evolved.

An alternative hypothesis suggested that prides evolved for cub protection; groups of lionesses that communally reared their cubs had more surviving offspring and therefore higher reproductive fitness. Indeed, females often come into oestrous at the same time, so give birth in synchrony. Mothers with similarly aged cubs form a 'crèche' and will nurse the offspring of their relatives, which does confer real benefits when it comes to cub defence, because groups of mothers have a chance at fending off an infanticidal male, whereas a lone female does not. Where this explanation fails is the fact that most felids, including leopards, tigers, puma and lynx, are also infanticidal, but remain solitary. Again, the fact that lionesses do cooperate to rear cubs together is not the same thing as group-living having evolved to combat infanticide.

For the Serengeti lions, the data showed that cooperative partners had the biggest impact on territorial defence. This is because, as already mentioned, a lion's worst enemy is another lion.* When different prides meet fur will fly, particularly if one group is challenging to steal another's territory. The African savannah is hugely diverse and, like a Monopoly board, certain areas are much more desirable than others. The important features for any lion territory are prey, cover to stalk that prey and provide vital shade, and a reliable water supply. Packer and team have been able to construct a map of lion 'real estate' in the Serengeti and it turns out the prime hotspots – the Park Lanes and Mayfairs of the lion world – are river confluences. They tick all the boxes, providing water, shade, cover and, during migration, a bonanza of zebra and wildebeest that get funnelled in as they try to cross the river. The data show that cubs born to prides that hold river confluences are more likely to survive to independence than those of other prides, which translates into a strong evolutionary pressure for territorial defence. It's much easier to defend a territory when you've got buddies to rely on, so bigger groups are able to defend bigger and better territories. Up to a limit, anyway. Groups can't keep getting bigger because more fighters also means more mouths to feed, so at some point the reduced food intake per lion caps the optimal group size for that particular patch. But the data support this hypothesis better than any other. In the Serengeti, at least, group-living seems to have evolved for securing land.

One query is why leopards have remained solitary, since they share the same environment as lions and need the same things. It's not been definitively answered, but what is known is that lions and leopards are completely intolerant of each

---

*Evolutionarily speaking. As with wolves, this is probably still true within national parks, but outside park boundaries a lion's worst enemy is most likely a human.

other. Lions will go out of their way to attack and kill leopards in the area, and leopards, for their part, will kill and eat lion cubs. It seems like there's only room for one social big cat, so leopards took a different evolutionary trajectory, becoming stealth machines that slip through the savannah unnoticed. Lions' aggression is not specific to leopards either: they are responsible for about 50 per cent of all cheetah cub deaths and about 30 per cent of wild dog pup deaths. They rule over the African savannah with an iron fist.

Although territorial defence is a huge part of lion life, studying it was not altogether straightforward. Territorial intrusions are rare and unpredictable, which isn't a good thing for a research project. What's more, much of the action happens overnight when visibility is poor. Instead of trying to observe it in action, in the early 2000s the team trialled an experimental approach, using loudspeakers to broadcast the roars of unfamiliar lions. The trials were led by Karen McComb, now professor of animal behaviour and cognition at the University of Sussex, and they worked a treat: the lions were completely fooled by the recordings and the way that they responded provided some valuable insights. A key finding was that lions' responses to simulated intruders depended on their situation. When the roars of unfamiliar males were played to groups of resident male coalitions, they all immediately responded, roaring back at the 'intruders' and approaching the speaker. Clearly, they were willing to engage in conflict. When the same roars were played to nomadic male coalitions, none of them roared back, nor did they approach the loudspeaker; in fact, half of them moved away from it. But once a nomadic coalition took over a pride, they behaved as the other resident males. Males are only willing to fight, it seems, when they have something to lose. For pride holders, the potential cost of being deposed is so significant that they don't have a choice.

McComb also evaluated how females responded to different roars. She found that lionesses with cubs remained relaxed when they heard the roars of resident males (i.e. their cubs' fathers), but became immediately agitated and rapidly retreated with their cubs when the roars of unfamiliar (and therefore potentially infanticidal) males were played. When female roars were played, resident lionesses immediately moved towards the speaker when the 'intruder' was a single female, irrespective of whether they were alone or not. When, however, the roars of three unfamiliar lionesses were played, resident females' responses depended on their own situation. Single females were cautious about approaching, whereas lionesses in groups of three or more showed less hesitation. What's more, if some members of their group were absent at the time, females often roared to them, recruiting them to help deal with the incursion. It reveals a flexibility in their behaviour, with potentially some ability to assess number, which may reflect a strong selective pressure for avoiding costly conflict with larger groups.

Being fooled by playback experiments is one thing – and most of us would be if we heard a stranger's voice in our house. But what about a dummy human? Maybe, if it was the right angle, the right light. Lions, you may think, as mammals with keen senses of smell, surely wouldn't be fooled by a dummy lion. Think again.

The experiments were set up by Packer and his then-student, Peyton West, to investigate the function of lions' manes. The mane is a feature that has long puzzled biologists, and although Darwin had not studied the animals, he was among the first to suggest an answer: 'The mane of the lion forms a good defence against the one danger to which he is liable, namely the attacks of rival lions.' Unsupported by any evidence, this hypothesis was accepted for the next hundred or so years, until Schaller revisited it and suggested an alternative, signalling hypothesis. Specifically, Schaller proposed that manes may function to tell females something about male quality. This is known as a sexually selected trait, because it functions to enhance reproductive success, whereas a naturally selected trait functions to enhance survival (in certain cases, such as the peacock's magnificent tail, what is sexually selected is not always what is best for survival). Schaller's proposal hinged on the fact that manes tick three key characteristics of sexually selected traits: they're found only in one sex,[*] they develop when the animal reaches breeding age, and they're highly variable – both between individuals (in colour, from almost white to black and in size, from a 'Mohawk-like' strip to a plush covering of the shoulders and chest) and within a lion's lifetime.[†]

The protection hypothesis was plausible: lethal claws and teeth mean that a 'shield' around the neck could be really beneficial. Nonetheless, West found no evidence for its two key predictions: that the neck is attacked more than other parts of the body and that wounds to the neck are more likely to be fatal than to other places. Instead, West's data fitted much better with Schaller's signalling hypothesis. For starters, darker-maned lions have higher levels of the hormone testosterone than their blonde-maned counterparts.

---

[*] To Schaller's knowledge – we now know that maned lionesses do exist and they demonstrate some behavioural characteristics of males. The phenomenon, which is rare, is thought to result from increased testosterone levels as the lionesses mature and these individuals are usually infertile.

[†] At a population level, average mane length varies by location too: males in Kenya's Tsavo National Park, for example, sport a little blonde crest and some tufts around the face – nothing like the plush, flowing mane of North Africa's Barbary lions, or those sported by some lions in the Serengeti. It comes down to climate – Tsavo is considerably more arid than other regions and thick manes are a hindrance here. But take a Tsavo male to a cooler, wetter region and he will grow a longer mane.

Testosterone is associated with aggression and darker-maned males spent longer in control of a pride compared with light-maned males, as well as being generally better fed throughout the year. What's more, the offspring of dark-maned males were more likely to reach independence. Mane colour, West and Packer hypothesised, could be an honest signal of male quality, communicating to females something about that lion's ability to fight and, crucially, protect her cubs. Mane length may also be an honest signal – numerous reports exist of manes being drastically reduced and even falling out following injury. Length may therefore provide useful information to males and females about recent fighting success.

Here's where the dummy lions came in. To test whether lions paid any attention to mane characteristics, West and Packer commissioned a Dutch toy company to build four life-size model lions, identical but for mane colour and length. 'Romeo' had a short, dark mane; 'Lothario' had a short, blonde mane; 'Julio' had a long, dark mane; and 'Fabio' had a long, blonde mane. Different pairs of the models were placed out on the plains some distance from a male or female and their responses were noted. They found that males were most likely to approach (and therefore challenge) Lothario, the blonde, short-maned model, while females showed a clear preference for Romeo and Julio, the darker-maned models. The lionesses, by the way, were totally fooled by the models; they are said to have slinked and rolled around provocatively next to their chosen mate, doing everything they could to show that they were interested. Up to the point where they sniffed under their tails. Packer has commented that if a lion could look affronted this would be it. Once fooled, but never again: only one trial could be conducted with each female.

If females prefer darker-maned males and males are less likely to challenge those with long, dark manes, then why don't all lions look like Julio? The answer is that dark manes come with a cost: in this case, extra heat. Using an infrared camera, West found that darker-maned males had significantly higher body temperatures than those with light manes. Lions are extremely susceptible to heat stress and darker manes are genuine handicaps because not only do they absorb more energy from the sun, they are also thicker than light manes so trap more heat. Only the strongest males can sport a dark mane and survive, making it an honest signal of quality. What this means, rather counter-intuitively, is that the most attractive, highest quality lions are the ones that need to laze in the shade for most of the day; they are the ones that are less capable of expending energy hunting. These males are built for fighting and sex only.

To minimise the risks of heat stress, lions are most active after dusk. Peak hunting takes place around dawn, so 'our' pride in the Delta was likely returning from a hunt and probably already enjoying a good post-prandial snooze. Lions can be inactive for up to 20 hours of each day, a fact perfectly articulated by Packer and Pusey when they wrote: 'Lions are supremely adept at doing nothing. To the list of inert noble gases, including krypton, argon and neon, we would add lion.'

Lions do better together because it enables them to defend good territories, and that in turn increases the chance that their cubs will survive and go on to reproduce. But within the group, how cooperative are they with each other and what does this tell us about the workings of their minds?

A great deal of research has been done to evaluate the conditions under which cooperation could evolve, much of it in an area called 'evolutionary game theory'. In essence, this is the use of theoretical 'games' to mathematically model social interactions and their outcomes; by manipulating different factors (such as how cooperative each individual is to the other, or how costly the cooperation is), you can predict the circumstances under which stable cooperation could evolve. One of the best-known models is called the 'prisoner's dilemma' and it goes like this. Imagine that you and your partner have just robbed a bank, but you both got caught and are now in jail being separately interrogated. The inspector outlines your options. Say nothing and, if your partner also says nothing, you'll both get two years in jail; but if you say nothing and your partner dobs you in, he'll go free while you get 10 years. Conversely, if your partner keeps schtum and you dob him in, you'll go free and he will get 10 years. If, however, you both tell on each other you will both get five years. What do you do?

The lowest jail time for both of you is if you mutually cooperate and say nothing. The big problem is that you don't know what your partner will do; if he cheats on you then saying nothing turns into the worst option. That's a big risk. If you forget about your partner and just think about yourself, then your best option is to cheat. Scientists working on simulations of these models have found that in a 'one-shot' game of the prisoner's dilemma, cheating comes out on top because you either get no jail time or you get equal jail time with your partner; you never do worse than him. Translating theory into reality, a version of the prisoner's dilemma was played out for real in the late-2000s British game show *Golden Balls* – it provided a devastating insight into the complexities of human nature. In the final round of *Golden Balls*, each of the two contestants was in the running for a big cash prize. Independently, they had to decide whether to split the prize with their co-contestant or steal it. If both chose to split, then both walked away with half the prize money. If one chose to split and the other chose to steal, the thief got the entire prize and the sharer got nothing. And if both chose to steal, both got nothing. Pretty brutal, and so were some of the show's contestants. Because the other thing about *Golden Balls*, which differs to the basic prisoner's dilemma game, is that the contestants could talk to each other before they made their final decisions. Here was their opportunity to reassure each other of their honesty and trustworthiness, to look the other in the eye and say, 'I promise I will split it with you.'

In one, infamous, episode, the finalists duly promised each other they would split the jackpot of over £100,000. He took her hand and told her that there was no way he would steal; and she looked him in the eye and said that her friends would be disgusted if she stole the prize. Clearly, they were going to cooperate and split the prize, it was a done deal. Until it wasn't. When the big reveal came, he split and she stole, and in that moment you saw just how well we humans have mastered the art of social manipulation. In a one-shot interaction with a stranger, it pays to cheat.

In the natural world, one-shot interactions with strangers aren't the norm. Within a lion pride, wolf pack or baboon troop, individuals are often related, but even if they aren't they recognise each other and repeatedly interact. Cheats cannot simply walk away and this changes things, because not only does the cheat leave itself susceptible to punishment, the wronged individual can use the experience to decide whether to cooperate in future. Now, the game becomes an 'iterated' prisoner's dilemma. In *Golden Balls*, the optimal way to behave would change if the finalists had to play several games against each other, because they would start to accumulate past form. They'd get a reputation. It's unlikely that a contestant that was cheated on in the first game would trust the other person again, likely leading to a string of games where both players cheated and neither received any prize. That means cheating might not be the most profitable strategy over the long term. It wouldn't take many games to be played before the amount the cheater won in the first game was less than the amount the pair could win by cooperating and splitting the prize every time. The opponent in this case turns into a partner. Hence, in social groups where individuals repeatedly interact with each other, cooperation could evolve in the form of a 'tit for tat' strategy where 'I'll scratch your back if, and only if, you scratch mine at a later date.' In other words, reciprocity.

Given the amount of cooperative behaviour they exhibit, coupled with the fact that they clearly recognise and differentiate pride members from unfamiliar animals, lions offered a good system in which to look for reciprocity. Packer and Robert Heinsohn did so with a focus on lioness territorial defence behaviour, using a similar experimental set-up to McComb to evaluate how they responded to simulated territorial intrusions. From observations with eight prides in the Serengeti and Ngorongoro national parks, they found something quite striking about lioness behaviour. They were not fully cooperative. Some individuals ('leaders') always led the charge to the perceived intruders, while others ('laggards') were consistently behind, arriving half a minute to two minutes later. Since territorial fights can be deadly, this puts the leader at a distinct disadvantage. Indeed, it was reported that in five real territorial disputes where laggards were apparent, the victim of the fight was always the lead female. Some of the laggards upped their speed when the leader was outnumbered by simulated intruders, showing that they will cooperate when necessary but overall, the study indicated a decent amount of freeloading among lionesses. There was no evidence that laggards differed in their age, body size or relatedness to the leader, nor that these females 'made up' for their reticence during territorial defence by taking a greater role during hunting or nursing. Bizarrely, leaders seemed to be aware of the freeloaders: when paired with a laggard they were slower to approach the speaker and they stopped more often to look back at their lingering companion than when they were paired with another leader. And yet, in contrast to predictions from theories such as the prisoner's dilemma, there was no punishment, no attempt to enforce cooperation. What's going on?

'I've since come to re-evaluate the leader-laggard paper,' Packer writes to me. 'Those experiments only considered the initial approach to a tape-recorded stranger, but in the latter half of actual approaches to real trespassers, everyone works together to surround and attack the invader. So, it's not like one lion does a favour to another and then gets rewarded at a later time – instead they either work together to solve a problem or no one does anything.'

The situation is a little different for male lions. Historically, coalitions were assumed to be relatives and this explained the low levels of aggression seen between partners. In one of the first applications of DNA fingerprinting with mammals,[*] Packer and Pusey found otherwise. Their analysis revealed that 42 per cent of coalitions contained non-relatives; however, contrary to predictions, the relatedness of coalition partners had no impact on how the lions behaved to each other. Unrelated partners did not challenge consorting males any more than if they were related, and fights between males over females were no more common among unrelated males. For a male lion, the cost of being alone is so high that it pays for him to join up with other males irrespective of their relationship, although this only holds to a certain point. Unrelated males tend to form pairs or trios, whereas larger coalitions (up to 10 males have been reported) almost certainly comprise relatives. It comes back to inclusive fitness. A large coalition is a fearsome prospect for other males, because there are high odds of them wresting control from a smaller

---

[*]With this technique, blood is taken and enzymes are used to cut the DNA into pieces. The pattern of DNA pieces is visualised on a gel as a read out of bands, a little like a barcode. Relatedness is correlated with the number of bands that are shared.

coalition and defending their status as pride holders for a long time. The flip side is that not all of these males can breed, so some males shoulder the costs of fighting without any of the benefits. Except, when the males that are breeding are your brothers, you do benefit, albeit indirectly, because copies of your shared genes are passed on to their offspring. A large coalition of unrelated males, on the other hand, makes no evolutionary sense because no male would forego breeding for a non-relative. Unrelated coalitions usually number two or three, because this is the optimal number to be able to put up a good fight, while maximising the number of mating opportunities. The lion world is full of trade-offs.

The fact that almost as many lion coalitions were formed of unrelated as related individuals provided the opportunity to test the different models predicting cooperation: did males in coalitions cooperate because of relatedness (i.e. kin selection), reciprocity ('you scratch my back and at a later point I'll scratch yours') or mutualism ('it's beneficial for each of us to behave together in this way')? To test these, territorial intrusions were again simulated by playing the roars of unfamiliar lions. Forty playback experiments were run with 15 resident male coalitions. If cooperation among males was accounted for by kin selection, then unrelated coalitions were predicted to approach the threat more slowly than related groups. This was not the case. If cooperation was reciprocal in nature, then coalition members should be attentive to the behaviour of their companions. This also didn't hold. Unrelated males seem to work together because it's mutually beneficial: each lion directly benefits from defending their territory and securing mating opportunities, and because the consequences of losing a pride are so significant there's no temptation to cheat on their partners.

How, then, do we sum up the cooperative behaviour of lions? Here's Packer again: 'We looked for evidence of reciprocity in a very strict sense, but we kept coming back to the same conclusion that lions are so mutualistic they help each other in dangerous situations, regardless. Lions are group territorial, and they need to work together in the common struggle against their neighbours, and the pressures to cooperate are so strong, that they go forward, regardless.'

★ ★ ★

The final piece of the story is how lion cooperation relates to intelligence, and this links back to our earlier discussions of social cognition (see Chapter 2). All animals, irrespective of their social system, need the cognitive abilities to keep track of their ecological and physical environment; for example, to learn and remember not to play with a porcupine, if you're a young lion. As we heard, animals that live in social groups are thought to face additional pressures, and this is the basis of Byrne and Whiten's Machiavellian intelligence (aka social intelligence) hypothesis: greater complexity in social groups led to the evolution of greater complexity in cognitive abilities. We heard already how, across primate species, neocortex size positively correlates with reports of tactical deception, and it has been proposed that reciprocity links with intelligence because bigger brains may help to keep track of social interactions. More on that later. Lion cooperation is governed by mutualism, not reciprocity; nonetheless, studies of captive individuals are revealing more about how they work together to solve problems.

It's fair to say that cats – and particularly big cats – are woefully understudied for cognition. As we saw already, it's only in the past 30 or so years that a concerted effort has gone into studying the cognitive abilities of animals other than primates. Lions, as highly social, cooperative group-living

animals, offer an intriguing non-primate species in which to
look for evidence of the Machiavellian intelligence hypothesis.
This was the starting point for Natalia Borrego, research
associate with the Lion Research Center at the University of
Minnesota and postdoctoral teaching fellow at the American
University in Cairo, who has been researching lion behaviour
for the past decade.

Borrego was interested in studying the evolution of
intelligence for her PhD and, rather than conduct
correlational studies of social complexity and behaviour, she
wanted to do experiments. Lions were an ideal study species
because their closest relatives (the other members of the
Panthera genus: leopards, tigers and jaguar) are all solitary.
Leopards also live in the same habitat as lions, while the
others are classed as having roughly similar ecological
complexity. As Borrego tells me: 'It's kind of as good as you
can get. That's why I picked lions, so I could compare them
to their asocial relatives and see if the social intelligence
hypothesis held.' On paper, it seemed like the ideal system. In
reality, other researchers weren't optimistic. 'When I said I
was going to test lions, everyone kind of said, "Uh, OK…"'
She laughs. 'Even Craig [Packer] at one point said to me,
"Sure, I'm fascinated by this but they're pretty lazy, so I don't
have high hopes."'

Undeterred, Borrego partnered with sanctuaries in Florida
and South Africa to conduct problem-solving tests with
captive animals. Her main challenge turned out to be practical
– lions are, perhaps unsurprisingly, incredibly good at
destroying things, meaning that, if the apparatus can't be
lion-proofed, it's at least got to be easy to put back together.
Patience and perseverance are also important, as Borrego
explains: 'One pair of lions stole my puzzle box for three days
and I could not get it back … Studying them is equal parts
fun and equal parts frustrating.'

Borrego has conducted the first test of cooperative problem-solving in lions. Using a variant of the same rope-pulling task, as described earlier with dogs and wolves, she paired up captive lions and observed whether they would work together to bring in the platform. Five of the seven pairs could, with four of them pulling together on their first trial. The more tolerant the lions were towards their partner, the better they did – the few pairs that squabbled performed worse at the task. For the most part, however, the pairs were remarkably harmonious, as predicted by their egalitarian social structure. And while they were able to pull together to succeed on the task, unlike the dogs and wolves, the one thing they didn't seem to do was coordinate their actions; for the most part, each lion seemed to just come in and get on with their bit of the task, rather than actively paying attention to what their partner was doing. Experiments in which one of the partners is delayed will help to tease apart whether lions can coordinate their actions by waiting for the other to arrive, but so far this fits with their status as mutualistic cooperators. In the wild, coordinated hunting is only known for lions living in the semi-arid environment of Etosha National Park in Namibia: here, lionesses work as a team when they hunt, with some individuals preferring to be 'wingers' that circle around the prey and some preferring to be 'centres' that lie in wait to ambush fleeing animals. It's thought that the reduced prey density at Etosha has favoured coordinated hunting because in this harsher landscape, solitary hunts are far less successful than in the richer environments inhabited by east African lions.

Borrego also investigated the problem-solving ability of individual lions, setting them the challenge of extracting a chunk of raw steak from a puzzle box in their enclosure. To open the box, the animal needed to pull back on a rope that was attached to a spring-loaded gate. Eleven of twelve animals succeeded, six of them on their first trial. Subsequent trials were solved faster, showing the animals had learned the technique and they remembered it even after a seven-month delay. She also set the same challenge to captive leopards and tigers, as the lion's asocial close relatives, and to spotted hyaenas. Hyaenas are social carnivores that sit within their own family, the *Hyaenidae* – while they look more like dogs, they're actually evolutionarily closer to cats. They are an interesting comparator group because they're highly social, living in 'clans' that can number from 15 to 130 individuals, but the groups are also strictly hierarchical, unlike the egalitarian lion pride. Borrego found that most of the hyaenas and lions solved the problem with ease, whereas only half the leopards and a quarter of the tigers got it. Lions and hyaenas were also more persistent than the solitary cats, spending longer trying to get into the box on their first trial. On this type of task at least, it does seem that social species have the upper hand when it comes to solving problems.

Lions are undoubtedly one of the most intensely studied large carnivores, but there's still much more to be learned. The caveat to most of what we know, which Borrego is keen to emphasise, is that it is specific to lions of the Serengeti. This study has provided a classic model of how lion ecology and behaviour is supposed to work; however, other lions have different social systems. Groups are smaller in other parts of Botswana and in Tsavo national park in Kenya, and, as we just heard, the lions at Etosha seem to be more cooperative than other lions. Much less is known about the behaviour of Asiatic lions. One of Borrego's ongoing research questions is how this variation in social structure influences lion behaviour and cognition, but she also acknowledges that social complexity is likely not the be-all and end-all. 'There's an idea that it's either ecological complexity or social complexity,' she says of the evolution of intelligence. 'But in all likelihood it's a combination of both.'

The big problem, as for large carnivores across the board, is that time is running out. Lions have been lost from more than 95 per cent of their historic range and are today virtually extinct outside sub-Saharan Africa. In June 2020, it was estimated that just 674 Asiatic lions* live in the Gir Forest (an area smaller than Greater London) and all other lion subspecies have been lost. With 20,000 individuals in the wild, African lions are faring better than their Asiatic counterparts in terms of absolute numbers, but this is still a catastrophic decline from the 200,000 that were living in Africa just 100 years ago. Lions are now extinct in 26 African countries and on the brink of it in many of the 28 where they are found. Only six countries have more than 1,000 lions – these represent the countries with the biggest and best-managed protected areas.

Is it possible to imagine a world without lions? As one of the planet's most iconic animals, so firmly ingrained in mythology and folklore, it seems a ludicrous thought. But the increasing pressures of habitat loss, hunting and conflict with local people mean the lion's future is far from certain. Some lion experts predict the animal could be extinct from all but the best-managed African reserves by 2035. For all of us, this should be a devastating thought.

Fact or fiction?

According to the Aesopic fables and countless other stories since, the lion is a big, lone male that rules over the other beasts; it is a deadly but noble animal that returns a favour when helped. There are a couple of fallacies here. First, and most importantly, there is no 'alpha male' in lion society – at

---

*In fact, Asiatic lions represent a conservation success story: in 1907, there were just 13 animals left. The 2020 estimate is a 29 per cent increase on the 2015 census figure of 523 animals.

least for African lions:* 'The reason they're so cool is that they don't have a dominance hierarchy,' Borrego laughs, 'it's not like *The Lion King!*' They are capable problem-solvers and seem to have good memories but there's no evidence for reciprocity. Lion cooperation seems to boil down to helping relatives and mutual self-interest.

If not the lion, then which animal shows better evidence of reciprocity? That very much depends on who you ask. Some have argued that it must be rare in non-human animals, while others document its presence in a diverse range of animal groups. The distinction, as always, comes down to definitions and interpretation, as Gerry Carter, assistant professor at the Ohio State University, USA, explains: 'Between the 1980s and 2010, the scientific consensus had shifted from "reciprocity is very important and common" to "it's very rare". It's still controversial, but it's actually a semantic disagreement.'

A big factor that has led many to believe that reciprocity in animals must be rare is the assumption of associated cognition, as Manon Schweinfurth, lecturer in psychology at the University of St Andrews, Scotland, tells me: 'Some people assume that reciprocity must be highly cognitively challenging. That you need some kind of mental notebook for recording interactions.' According to this view, she explains, stable cooperation could only evolve if animals could keep track of others' behaviour and punish freeloaders, as we do. Schweinfurth, Carter and others take a less restrictive view, proposing instead that reciprocity is an umbrella term for several different kinds of interactions, which can be achieved in very different ways. Hermaphroditic flatworms, for example, take turns to be the male during

---

*Things look a little different for Asiatic lions – there is evidence for dominance hierarchies within male coalitions and, most recently, studies of females also suggest within-pride hierarchies.

mating. These worms possess both male and female genitalia, but there's a stronger incentive for both to be the male because that way they can sire more offspring. This leads to conflict. They resolve this by taking it in turns to release an egg that the other fertilises, something that meets the general definition of reciprocity, but requires no cognition to achieve. 'They're reciprocating beautifully, and they're worms,' says Schweinfurth. 'They don't even have a brain!'

In contrast, our big brains have enabled extreme levels of cooperation that do involve cognition. Individuals go out of their way to donate money, time or even blood to strangers that they may never meet, let alone see again, and entire nations can cooperate with each other. We also help friends and colleagues that we see more regularly and in these relationships there's usually an expectation that favours will be returned. The paradox of reciprocity is that it's susceptible to cheats, to freeloaders that accept help but either don't pay it back or pay back less than they could. We overcome this through several mechanisms that help to enforce cooperation and it's been argued that the number of mechanisms we use enabled higher levels of cooperation to evolve and stabilise among unrelated groups. We reward those that cooperate with praise or reputation-building, which makes the cooperator feel good and probably increases the likelihood of them doing the same again. We also 'punish' non-cooperators by withdrawing our help to them, encouraging others not to partner with them or excluding that person from social activities. One study found that people were actually willing to pay money to punish cheats, demonstrating just how fundamental cooperation is in our societies.

Those abilities don't come cheap and that led some, including Trivers, to speculate whether the importance of altruism in human evolution was selection for the evolution of advanced cognition, resulting in an increased brain size.

We're back to the social intelligence hypothesis again. On the face of it, it certainly seems as if reciprocating with others would require some impressive cognitive feats: you have to remember not only who you have interacted with, but how much of a particular commodity you owe them or they owe you, and potentially what that would be worth if one of you wants to pay it back in a different commodity. You then have to scale that up for every individual you regularly interact with. That's exhausting enough to just think through, so can we really believe that other animals do the same? Schweinfurth smiles and shakes her head: 'there are other cognitive mechanisms that are way easier.'

Many examples of reciprocity have been reported in the animal kingdom, but to understand more about how they do it, we need to look at the finer details. Fortunately, there are a couple of study systems that are providing some fascinating insights. And, you may be surprised to hear, none of them are in apes or corvids.

We start with the classic example of animal reciprocity, which takes us to the rainforests of Costa Rica in the late 1970s. Student Gerald Wilkinson was conducting his PhD research on vampire bats, which are the only known mammals to exclusively feed on blood (called 'hematophagy'). They usually feed on horses or cattle, which they crawl to at night on all fours; using their razor-sharp teeth they puncture the skin and lap up the blood, depositing a cocktail of anticoagulant salivary proteins[*] into the wound to prevent it from clotting. Feeding can take anywhere from 10 minutes to one hour and during this time the bat can drink more than its body weight in blood, swelling up like a huge, furry mosquito. In one year, a colony of 100 bats can drink the blood of 25 cows, which is the sort of media headline that makes everybody

---

[*] One of the proteins is called draculin – I kid you not.

panic, but break it down to how much each bat consumes each night and it's not enough to do any damage.

The problem for vampire bats is that they don't always find food. From his observations of 184 individually tagged bats, Wilkinson estimated that almost one in 10 adults and one in three juveniles come home hungry every night. A bat that fails to feed for three consecutive nights will almost certainly die and this is where reciprocity comes in. Wilkinson recorded hungry bats in the roost begging to well-fed bats and the latter regurgitating congealed blood to them. Most of the reports were of mothers sharing with their hungry offspring, but not all. Sometimes, unrelated individuals shared blood meals too and these were individuals that frequently associated together.

Wilkinson didn't immediately realise the significance of what he had found, commenting in a later interview that: 'It was only afterwards that we said – "Oh my goodness! Look at this!" And it was only at that point that I thought – "Wow! This is actually pretty cool."' He published his findings in 1984 and they generated a huge buzz. Trivers interviewed Wilkinson in depth and featured the research in his 1985 book, *Social Evolution*; as a result, blood sharing by vampire bats became the definitive example of reciprocity in the animal kingdom.

Not everyone agreed, of course. Critics noted the low number of unrelated individuals, dismissing the claim of reciprocity as reflecting anomalies in an otherwise kin-biased sharing system. Wilkinson set it to one side, but in 2013 was approached by Carter, who became interested in food sharing while studying for his master's: 'I started reading the literature on that topic, and realised there was a lot of controversy and modelling, but no one was actually collecting more data or doing more experiments.' Carter conducted his PhD research with Wilkinson at the University of Maryland, USA, replicating the original result and carrying out further experiments to address some of the critiques.

One of his studies with 20 captive bats found that 64 per cent of sharing partners were unrelated and that the strongest predictor of whether a bat donated blood was whether it had previously received a donation – this was over eight times more important than relatedness in predicting food-sharing. The partnerships were stable, indicating that these little bats had formed long-term social bonds. In fact, Carter has been able to see this in action by housing complete strangers together in captivity. As he tells me: 'Realising we can watch how new relationships form is the most exciting recent thing for me. Another big one is our ability to track relationships using these amazing new proximity loggers. It is surprising how well they work. We can track bat relationships from captivity to the wild and back.' The biologgers can weigh as little as 1g and look set to provide unrivalled data about the bats' social relationships. Carter's ongoing work is likely to shed much more light on the evolution of cooperation in vampire bats.

The bat studies suggest that reciprocity has an emotional basis, since bats preferentially shared with 'friends'; i.e. long-term socially bonded partners. This is one form of reciprocity, but not the only one, as is apparent from a series of clever studies with rats.

Schweinfurth investigated rat reciprocity for her PhD at the University of Bern, Switzerland, in the lab of behavioural ecologist Michael Taborsky. Taborsky had established model systems with Norway rats and cichlid fish to investigate cooperative behaviour, enabling an experimental approach to be taken, something that appealed to Schweinfurth's interest in understanding not just can the rats reciprocate but how.

Rats are much-maligned little animals, but in fact they are sociable creatures that go out of their way to cooperate. If, for example, rats have the choice to get food by either working on a cooperative task with another rat or doing a task on their own for the same reward, they prefer the sociable task.

They show different kinds of reciprocity. Females, for example, are more likely to help a new partner when they've recently received help from a stranger than rats that hadn't received any help. This is called generalised reciprocity and it's akin to the 'pay it forward' kind of helping we do if someone's gone out of their way to help us and there's no opportunity to pay that person back.

Male and female rats both show direct reciprocity, which is the more intuitive sense of the term – when you help a particular individual that has previously helped you. Demonstrating this experimentally involved setting pairs of rats a version of the classic rope-pulling task. Food was placed on a little, out-of-reach platform. Attached to the platform was a handle, which poked into the cage of one of the rats. By pulling the handle, the platform moved into the cage and the food became accessible – but only to the partner rat. Thereafter, the roles were exchanged. Now the partner rat had the opportunity to reciprocate the favour by providing food to the previously cooperative rat. By manipulating the identity and helpfulness of the social partner, the team found that rats were more likely to donate food to a partner who had previously helped them as compared with a non-helpful partner. What's more, they took note of the quality of help they received, adjusting the help they provided that individual accordingly. Rats are also sensitive to their partner's need and pulled in the platform more often when their partner was hungry and weak. Recent evidence suggests that rats can detect genuine need from fakery (another form of freeloading). Some clever experimental manipulations allowed Taborsky's team to show that when rats could 'smell hunger' they were much quicker to provision a partner with food than when they could smell a satiated rat. Smell may therefore act as an honest signal of others' need.

These findings show that rats base their behaviour on partner-specific information, like the vampire bats. But there's a big difference: the rats' decisions are based simply on their most recent interaction with a particular rat, rather than any deeper social bond. In the world of the rat, a partner could have cooperated on 100 previous trials, but none of that matters if, on trial 101, they cheat: a rat will cheat on them in return. It seems a little harsh to us (surely, we should all forgive an occasional slip-up?), but comes down to the fact that rats are processing information about their social group in a very different way. Rats are generally nice to each other, Schweinfurth explains, but they seem not to form long-term social bonds like vampire bats or humans, something that she found astonishing: 'I think this sets apart how rats reciprocate from how we reciprocate. We accumulate knowledge and form a bond with someone that helped us repeatedly. Rats seem to be constantly overriding information; they only keep track of their last experience and that's cognitively less challenging.' It's a different kind of reciprocity, known as 'attitudinal' and for Schweinfurth it emphasises the importance of reciprocity for rats: their use of a strict tit-for-tat strategy achieves cooperation efficiently, without the need for social bonds or advanced cognition. Carter agrees: 'It's very clear to me that what the rats are doing in those experiments is different than what the vampire bats are doing. The rats are cooperating based on recent actions only, not on long-term relationships.'

The scientific reception to Schweinfurth and Taborsky's findings was mixed, as she tells me: 'People would say, "Oh, yes, your rats seem to be able to reciprocate, but they're probably an exception."' At the root of this was the fact that chimpanzees and other apes were thought to be incapable of reciprocity. If they couldn't do it, then its presence in rats and bats must be down to some odd quirk of evolution. Schweinfurth was not convinced so, together with psychologist Josep Call at St Andrews, she conducted an in–depth review of all papers ever published on reciprocity in primates. They found that for observational and experimental data, there were actually more studies showing evidence for reciprocity than studies showing no evidence. Their analysis suggests that reciprocity didn't evolve in the human lineage, with rats and bats as weird outliers. Schweinfurth continues to work with rats, but is now also working on chimpanzees. She wants to understand how comparable ape reciprocity is to that exhibited by humans.

The best evidence for reciprocity in primates is in capuchin monkeys, with much of this coming from the lab of primatologist Frans de Waal at Emory University's Living Links Center in Georgia, USA. De Waal, now emeritus C. H. Candler professor of primate behaviour at Emory, found evidence for reciprocity in captive brown capuchins, using several different tests. For example, the monkeys will spontaneously hand food to monkeys in neighbouring compartments and the recipients will return the favour. Capuchins have demonstrated both generalised and direct reciprocity, paying forward positive and negative outcomes using a simple 'give what you received' rule. They've also shown evidence for attitudinal and emotion-based reciprocity, suggesting that they may flexibly use both depending on the context, just as we do. This is apparent from observations showing that they will reciprocate with familiar and unfamiliar partners in the short term (like the rats), but that they also form longer-term bonds with individuals. One experiment directly tested the two alternatives and found that monkeys were more likely to share food with a 'friend' than with an unbonded partner that had just shared with them, indicating that friends trump helpful strangers, but that if no friends are around, they still know how to play tit-for-tat.

An equally fascinating finding to come out of the capuchin research is their perception of fairness. This is a fundamental trait in human societies; we have a very strong sense of what is fair and what is not, and we draw upon that to detect cheats. Finding out that someone else is being paid more to do the same role, for example, elicits feelings of outrage, prompting the intriguing question of whether other animals possess the same.

Sarah Brosnan, as part of her PhD research with de Waal, conducted one of the first tests of this question (known as 'inequity aversion') using capuchin monkeys. In this pioneering experiment, female* monkeys learned to exchange a small rock with a human experimenter for a piece of cucumber. Each trial involved two monkeys, which the experimenter alternated between; their cages were next to each other so they could see but not interact with each other. This was the straightforward part of the task: the monkeys easily learned the set-up and quickly exchanged the rock for cucumber on every trial. Brosnan then introduced inequality. Now, one monkey continued to receive cucumber each time she handed over the rock, but the other received a grape — which for monkeys is a considerably better reward. This did not go down well. It's hard not to anthropomorphise, but the unfairly rewarded monkeys started to look annoyed. One capuchin, continuing to receive cucumber while her neighbour was getting a grape, started pacing her cage and vocalising. Then, in an apparent fit of petulance, she started throwing the cucumber back at the experimenter or straight on to the floor. When the partner was given a grape 'for free' (i.e. without having to do anything), the reaction was even stronger. While we don't have access to what was going on in

---

*Pre-testing trials involved males and females, but the males were considerably less affected by inequity than the females, so the main experiment focused only on females.

the monkey's mind at this point, it seems that she was emotionally affected by the inequality in her pay-off.

Inequity aversion was subsequently found in chimpanzees and long-tailed macaques, but not in orangutans and squirrel monkeys, the latter two species being less naturally cooperative. Ravens and crows were also tested on the same kind of experiment: for them, grapes were the 'booby prize', compared with their favourite pieces of cheese. Like the monkeys, the corvids' behaviour indicated that they knew they were being unfairly treated – in this case, they stopped cooperating altogether, no longer exchanging the token for the reward.

While most evidence for inequity aversion comes from the negative reactions of disadvantaged individuals when they see another receive a bigger or better reward, some animals are also sensitive to being *over*-rewarded compared with a partner. Brosnan and de Waal found that in a replica of the capuchin task with chimpanzees, the animals reacted not just when they were disadvantaged but when their partner was too. In these situations, some individuals refused to accept the reward until their partner was offered the same. It's another way to safeguard cooperation – in the long term, the costs of your partner defecting because they weren't being treated fairly may be higher than the immediate cost to you of rejecting your reward.

The ability to detect unfairness may be a factor that humans use to enforce reciprocity, but inequity aversion is not necessary in order to reciprocate. Schweinfurth explains that the concepts are tightly linked: children develop a normative understanding that they should reciprocate from the age of three or four years, and this is about the same time that they become aware of inequity. They are likely also linked in chimpanzees, but they certainly don't need to be for all animals.

In Aesop's fable, the lion remembered its interaction with the shepherd and repaid the favour when the shepherd was in need. The lion and the shepherd were not friends, they didn't have a long-term social relationship. And because of that, our replacement for the lion in this fable has to be the rat. They are helpful, prosocial little animals that reciprocate based simply on their last interaction with an individual and an assessment of genuine need. That's what this story is all about, so 'The Rat and the Shepherd' is a better biological fit.

CHAPTER SEVEN

The Monkey and the Fisherman

A Monkey perched upon a lofty tree saw some Fishermen casting their nets into a river, and narrowly watched their proceedings. The Fishermen after a while gave up fishing, and on going home to dinner left their nets upon the bank. The Monkey, who is the most imitative of animals, descended from the treetop and endeavoured to do as they had done. Having handled the net, he threw it into the river, but became tangled in the meshes and drowned. With his last breath he said to himself, 'I am rightly served; for what business had I who had never handled a net to try and catch fish?'

It's 1948, and Kyoto University students Shunzo Kawamura and Jun'ichiro Itani are collecting data on a group of semi-wild horses at the Toi peninsula in the Miyazaki prefecture. The study forms part of a bigger research programme on social behaviour and evolution established by pioneering zoologist Kinji Imanishi. During the Second World War, Imanishi was sent to Mongolia and while there he became fascinated by the social behaviour of wild horses. He began recording how the animals interacted, using the then-novel approach of identifying and naming every individual in the group, and with the ultimate goal of better understanding the evolution of human societies. Imanishi considered horses to be the best animal for studying animal societies, so when he returned to Japan his attention turned to those living on the Toi peninsula.

Kawamura and Itani are diligent observers, yet on this day something else has captured their attention. A group of wild monkeys, Japanese macaques, has appeared in the woods. They have solemn, bare pink faces surrounded by a muffle of thick, spiky fur, and their behaviour captivates the students. Upon hearing their account of the animals, Imanishi arranged a trip to the tiny island of Koshima, just off the coast of the Toi peninsula and where macaques were known to live. As Tetsuro Matsuzawa, a distinguished professor at Kyoto University, has reflected, this was a pivotal moment in the history of primatology research: 'Friday, 3 December 1948, this is the day that Japanese primatology began, this is the day they visited Koshima Island to see wild Japanese monkeys.'

Imanishi was immediately impressed and research on the monkeys began soon after; it marked the beginning of a long-term continuous research effort that continues to this day. Initial observations were not easy, as intense hunting pressure meant that the animals were fearful of people and the thick evergreen forests offered limited visibility. Taking inspiration from birdwatching, the team started leaving food for the animals on the beach (in the form of sweet potatoes and unhusked wheat); as a result, they managed to draw the troop of 20 or so animals out into the open for easier observation. All that they needed to collect data was a pair of binoculars, a notebook and a pencil.

The sweet potatoes were often caked in mud, which the monkeys would brush off before eating. Then, in September 1953, a young female nicknamed Imo (which translates to 'potato') did something unusual. She took a muddy potato to a little freshwater stream near the shore and dunked it in the water, washing off the mud before eating it. She did the same the next time she had a muddy potato, and the next. Imo had come up with an innovative technique for processing the potatoes and in that first year her innovation spread to three other young monkeys. More followed suit: other youngsters first, then older siblings, parents and other adults. Old males were the least likely to adopt the new behaviour. After about five years, potato washing had become the norm in this troop and after this the direction of transmission changed: instead of adults learning it from young monkeys, it became one of the many behaviours that infants learned from their mothers.

Sweet potato washing is the iconic example of the Kyoto macaque research, although the spread of several other behaviours was subsequently reported, from Koshima and other study populations. It was the first published research indicating that innovations could be socially transmitted between members of an animal group and today no textbook of animal behaviour is complete without mention of Imo's innovation. The question is how the behaviour spread between individuals in the troop: Aesop painted the monkey as the most imitative of all animals, but does science agree?

The idea of monkeys as master mimics has a long history, and the fable of the monkey imitating the fisherman is not the only one of Aesop's to feature monkeys behaving this way. In 'The Dancing Monkeys', a troop of monkeys dressed like courtiers and mimicked the dancers beautifully; that is, until someone threw a handful of nuts on the floor and they reverted back to their natural behaviour. In fact, there's a striking parallel with this fable in Japan too. The ancient Japanese performing art of Sarumawashi quite literally translates as 'dancing monkeys'. It has evolved over a 1,000-year history, rooted in the fact that the monkeys were revered: the 'show' was considered more of a ceremony in which the monkeys acted as mediators between the audience and the gods. Traditionally, macaques were dressed in traditional warrior clothing, including the trademark hat, and trained to mimic the actions of humans in street-side performances.[*] It's not scientific, but it does show that monkeys can be trained to copy human behaviours, just as portrayed in Aesop's fable.

Although monkeys are not native to Greece, they nonetheless made popular pets and are well represented in writings and art in the seventh century BC, with intriguing evidence for other, more exotic, monkeys even further back.[†] It should be noted that in ancient Greece, and in

---

[*] It used to be a common street-side show, but has declined in popularity since the 1970s: a consequence of huge increases in motorised transport and a change in public opinion regarding the use of performing animals.
[†] In a building in Akrotiri, a settlement of the Minoan civilisation on the island of Thera (Santorini) that was buried by ash from a volcanic eruption around 1600 BC, several wall paintings show monkeys. Many of these are Egyptian species such as olive baboons, but one of them portrayed grey langur monkeys, which are native to Nepal, Bhutan and India. During the Bronze Age, this region also included the important Indus Valley Civilisation, so the fact that a Minoan artist painted grey langurs on a wall in Akrotiri suggests direct contact between these civilisations; something that as yet has not been conclusively proved.

many cultures since, there wasn't a distinction between monkey and ape: the same word was used to describe both, which means that many of the writings talk about 'apes'. Nonetheless, it is unlikely that Aesop came across chimpanzees, gorillas or orangutans; more likely he encountered monkeys imported from North Africa such as Barbary macaques or geladas. Some clues about which species were known come from Aristotle who, in his *History of Animals* summarised primate classification as: monkeys with no tails (Barbary macaque or ape), monkeys with tails and 'dog-headed' baboons.

The humanlike aspects of monkeys made them attractive characters for depicting humans; however, as in Aesop's fables, these were rarely flattering. Monkeys were used to represent dishonest, mean, ugly people; they were deserving subjects of scorn and mockery. In one fable, Jupiter holds a beauty competition for all the animals' babies and the monkey mother is met with derision for tenderly presenting her 'flat-nosed, hairless, ill-featured young Monkey as a candidate for the promised reward'. In other fables, monkeys suffer dire consequences for trying to be like people: the monkey that tried to fish paid for it with his life, while the dancing monkeys were ridiculed for reverting to type. Another fable tells of a ship's monkey which, after the vessel sank, tricked a dolphin into carrying it back to shore by pretending to be a man. When the dolphin discovered that it was only carrying a monkey, it drowned it. Monkeys, for Aesop and many writers since, were excellent imitators but, and it's an important point, this propensity for copying was equated with a lack of intelligence. Monkeys were mindless mimics, unable to think for themselves. We'll return to the significance of this shortly.

An article published in the *Wilmington Centinel*, North Carolina, USA, in 1789 described how, according to a person who had just returned from the Labrador coast:[*] 'The imitative faculty in monkeys seems to exceed everything short of human.' The piece described how a sailor went ashore to sell woollen caps and stopped for a rest under a tree, with one cap on his head and the others by his side. When he woke, he was astonished to find all his caps being worn by a group of monkeys in the tree, who were chattering excitedly to each other. The author described what happened next:

> Finding every attempt to regain them fruitless, he at length in a fit of rage and disappointment, and under the supposition the one he retained was not worth taking away, pulled the same from his head, and throwing it upon the ground exclaimed – 'here d—n you, take it among ye,' which he had no sooner done, than to his great surprise, the observant monkeys did the same, by which means he regained the greatest part of his property.[†]

Beliefs and stories about monkeys crystallised the notion of their mimicking prowess and even early scientific accounts accepted monkey mimicry as fact. Darwin's disciple, George Romanes, wrote in his book *Animal Intelligence* (1882) that:

> Allied, perhaps, to curiosity, and so connected with the emotions, is what Mr. Darwin calls 'the principle of imitation'. It is proverbial that monkeys carry this principle to ludicrous lengths, and they are the only animals which imitate for the mere sake of imitating...'

---

[*] Admittedly, the validity of the piece may be questionable, given how far away from monkey native habitat this would be.

[†] If it seems familiar, it's because the classic children's picture book, *Caps for Sale*, written and illustrated by Esphyr Slobodkina and published in 1940, tells the same story.

The science of animal imitation is more recent, though has a longer history than many of the topics in this book. Today, it is a vast field of research, involving primates, rats, birds, fish, cetaceans, reptiles, insects and others. A comprehensive overview is far beyond the scope of this chapter, indeed of a single book. Here, we stay largely focused on primates.

★ ★ ★

Primates is the evolutionary order that includes monkeys, together with tarsiers, lorises, bushbabies, lemurs, apes and humans. Within the primate group, there are several different lineages, but we're only going to consider the monkeys and apes (or 'anthropoids'). The ancestral anthropoids diverged in the region of 40 million years ago, when some of them made it across the Atlantic to South America and evolved into the New World monkeys (called *platyrrhini* for their broad, flat noses with sideways-facing nostrils). It was another 15 million years before the ancestral Old World anthropoids (called *catarrhini* for their narrow noses and downward-facing nostrils) diverged to form Old World monkeys and apes, with the latter subsequently splitting again, 15–20 million years ago, into the lesser apes (gibbons and siamangs) and great apes (chimpanzees, bonobos, gorillas, orangutans and humans; hereafter 'apes'). It's for this reason that the terms 'ape' and 'monkey' are used interchangeably and, in many languages, are the same word.

Old World monkeys include macaques, baboons, mandrills, langurs and other species native to Asia, Africa, the Middle East and Gibraltar. New World monkeys include howlers, capuchins, marmosets, tamarins, titi and spider monkeys, among others. There's no consensus on how the latter made it across the Atlantic — bizarre as it sounds, the frontrunner explanation is that they rafted across on mounds of vegetation — but once in the vast rainforests of Central and South America, they radiated into a spectacular diversity of species. They range in size from the world's tiniest monkey, the 100g pygmy marmoset, to the largest, the southern muriqi (aka southern woolly spider monkey) weighing 12–15kg. If you are lucky enough to have spent time in the Amazon rainforest, you will have heard howlers roaring at dawn (a sound not unlike a jet plane warming up its engines) and perhaps seen the regal splendour of emperor tamarins with their fantastic long whiskers.

Primates are an excellent group in which to learn more about social behaviour, because they exhibit a spectrum of social systems. At one extreme there are orangutans, which are solitary except for mothers with dependent offspring. There are then monogamous family groups (such as gibbons, titi monkeys or, mostly, humans), polyandrous family groups (where an additional male joins and helps with rearing offspring, such as marmosets and tamarins), polygynous groups (where one male controls a harem of breeding females, such as gorillas and hamadryas baboons) and, at the other extreme, larger groups with multiple males and females (such as Japanese macaques and savannah baboons). Across primates, there is much opportunity to learn from others and, spoiler alert, there's ample evidence that they do. But importantly, this is not the same as saying that they can all imitate. That's because once we start getting into the nitty-gritty of social learning, it becomes both complicated and contentious. As we saw already with animal intelligence, it's problematic to try and design experiments for something that has not been acceptably defined. And, unfortunately, terminological issues have plagued the field of animal social learning since it began.

★ ★ ★

For centuries, imitation has been a catch-all term applied to all sorts of phenomena in the animal world: from eye spots on butterflies to cuttlefish colour-matching their environments and lyre birds producing chainsaw sounds. It's also been used interchangeably with mimicry, with both words historically applied to any similarity of behaviour or appearance between two organisms. The word 'mimicry' derives from the Greek *mimetikos*, which means 'imitation' – hence the fluidity in use. Today, however, science distinguishes them. Mimicry is thought of as an evolved similarity in the appearance of one organism to a different species, often functioning as a dishonest signal to influence the behaviour of potential predators or prey. Imitation, on the other hand, comes under the category of behavioural similarity – we'll come back to its definition shortly.

There are many ways in which the behaviour of one animal can influence another and it's important to note that they don't all involve learning. Synchronicity is one example: shoaling fish and flocking birds can adjust their behaviour to match their near neighbour, leading to incredible collective behaviour involving thousands, if not millions, of individuals. Behavioural contagion is another: one animal's behaviour can automatically trigger production of the same in a bystander, which is thought to have evolved to help coordinate the actions of individuals in a social group. Yawning is a good example of contagious behaviour, something that may have once been beneficial if it encouraged all individuals in a group to rest at the same time.

Imitation, as a form of social learning, is distinct from these hardwired, stereotyped actions. That narrows things down a lot, but without further refinement still leaves the term covering a huge array of behaviours. In 1900, psychologist Conwy Lloyd Morgan summed up the problem – behaviours could be imitative in their *effects*, without necessarily being so in their *purpose*. With this he was largely railing against Romanes again – as you may recall from the discussions of intelligence, different animals can get to the same outcome but not necessarily in the same ways. It's an important distinction and Morgan proposed categorising imitation as either instinctive or reflective, with only reflective imitation revealing of intelligence.

Morgan's classification helped but it was still far too woolly for some. US psychologist Edward Thorndike, who founded the field of behaviourism, wrote in 1911: 'To the question, "Do animals imitate?" science has uniformly answered, "Yes." But so long as the question is left in this general form, no correct answer to it is possible.' The trouble was that psychologists were finding evidence that animals could learn from others, but what they saw didn't fit Morgan's neat categories. In one experiment by Melvin Haggerty, for example, two monkeys were presented with seven hanging strings, one of which needed to be pulled to release a treat. One monkey quickly learned the solution while the other never did, even after many trials. Yet, when Haggerty set up the task so that the 'stupid' monkey could watch the other solve the problem, it showed a clear change in behaviour:

> There had been ample opportunity for the second monkey to learn the trick unaided, but he had failed to do so; the strings had never brought satisfaction to him through his own activity. Yet now, although he did not use the strings to get food, he continued to handle them, to pound them against the side of the cage and against each other, and several times after acting in this way he looked directly into the food opening.

After another demonstration session, the second monkey focused its attention more on the correct string. This progressive narrowing of the animal's behaviour was clearly influenced by the demonstrating animal, but it was not an all-or-nothing process. Haggerty admitted that his data were inconclusive, but expressed dissatisfaction with the existing theory which, in his view, had reached its limit of usefulness.

For Bennett (Jeff) Galef, emeritus professor in psychology at McMaster University in Ontario, Canada, the long history of imitation research offers both 'comfort and confusion'. Comfort, Galef wrote in 1988, because it shows just how much interest there is in a topic that has stood the test of time. Confusion, on the other hand, because it's been approached and written about in such different ways that the resulting vocabulary was 'chaotic and contradictory'. Galef commented in a 2009 interview that the terminology problem was severe enough to have turned scientists away from the field; something that Haggerty had also noted in 1912.

Has it really been so bad? I put it to Andy Whiten, who we previously met in Chapter 2. Subsequently, Whiten's research interests shifted towards social learning, another adaptation for living in social groups. 'I agree with him [Galef],' Whiten responds. 'And it's still the case. There are two bases for disagreements over which species imitates – first, how imitation is defined; and second, for any definition of imitation, what is established empirically?'

The crux of the 'how' question is that imitation is not the only process by which the actions of one animal (we'll call it the 'demonstrator') can bias the learning of another (the 'observer'). In Thorndike's day these hadn't been clarified, leading him to conclude that imitation would be demonstrated if an animal could, 'from an act witnessed, learn to produce that act'. We now know that it is a lot more complicated – Thorndike's definition would be better applied to the umbrella term of 'social learning', which covers a spectrum of different processes associated with different cognitive requirements.

At the simpler end, there are motivational and perceptual processes that may bias an observer's learning when it sees another animal doing something. Motivationally, the mere presence of the demonstrator can reduce stress levels or increase overall arousal in an observer, increasing the likelihood of it engaging in a task and learning about it. Crows, which as we have heard are highly neophobic, are more likely to interact with a novel food when they see another crow feeding from it first, which is likely due to reductions in fear increasing the likelihood that it contacts the food and learns its value. This is called social facilitation and it is a hugely important process in the natural lives of animals.

Perceptually, an observer's attention can be drawn to what another animal is interacting with, encouraging the observer to independently explore that place or object and learn about it. This is called stimulus enhancement and it likely accounts for many examples of so-called imitation in the early literature, including Haggerty's monkeys. It's also thought to be the process involved in one of the most iconic examples of social learning in a wild population: blue tits raiding milk bottles in Britain. Here, milk used to be delivered in glass bottles that were capped with cardboard and, later, foil, which seems like it should be secure. Yet homeowners became perplexed when their bottles started being broken into and the wonderful thick layer of cream skimmed off before they brought them in. The culprits were blue tits who, along with robins, had reaped the rewards from milk bottles in the days when they were delivered with no cap at all. Robins were deterred by the addition of caps, blue tits weren't. The first record was in 1921 in southern England and by the 1940s it had spread across the whole of the country: far too quickly to be explained by independent individual discovery. Something was happening at a social level. And, just like Haggerty's monkeys, it wasn't that each bird was copying another, but their attention was drawn to the presence of other birds interacting with pierced bottle tops, just as the unsuccessful monkey's attention had been drawn to the strings. Once they had learned that tasty cream was available, there was an increased likelihood that they would peck at a sealed cap, discovering their own solution through trial and error.

Facilitation and enhancement both increase the probability that an animal will interact with a place or object in its environment, but that's as far as the social influence goes – further discovery and learning is usually accounted for by individual trial-and-error (i.e. 'asocial' learning). Although these processes may be of huge significance to wild animals, they've been largely sidelined in the quest for imitation – Galef has commented that finding evidence for this ability in animals came to be something of a 'holy grail' among psychologists. 'It's absurd isn't it?' This is Dick Byrne, who we also previously met in Chapter 2. 'And I think it's led to a real waste of time in animal psychology. Because people are just trying to get a tick box so they can say, "look my animal imitates", without thinking why.' This focus on imitation has led to a huge number of studies; however, summarising them is not straightforward because scientists have thought about imitation in staggeringly different ways: from its definition to its method of study and its interpretation. Not that this sets animal social learning apart from other areas of scientific study – debate and disagreement is normal. It's just that there has been (and continues to be) rather a lot of it in this field.

To control for the possibility that stimulus enhancement was responsible for apparently imitative learning, in the 1960s the 'two-action' task was born, initially for an experiment with budgerigars. By the 1980s it was the definitive method for laboratory studies of imitation. The two-action task makes use of cleverly designed pieces of apparatus, in which food can be extracted in either of two different ways. Importantly, both methods involve manipulation of the same part of the apparatus, such as a lever or door, which is how it controls for enhancement. Observers watch only one of the techniques and the question is whether they subsequently solve the task using the demonstrated technique. Early studies included rats watching demonstrators move a little joystick to the left or the right, and quail watching demonstrators peck at or step on a little pedal.

Two-action tasks may have enabled greater methodological control, but did they really reveal imitation? Two lines of reasoning argued not.

The first comes down to what the animal learns from a demonstration. For imitation, it needs to learn the form of the demonstrator's actions – what it's doing and in what order. Alternatively, psychologist Michael Tomasello suggested, the observer could learn something about the end result of the demonstration and how the objects move, subsequently attempting to reach the same result in its own way. He defined this in 1990 as 'emulation' learning and a subsequent study with kea provides a good example. These inquisitive New Zealand parrots observed a trained demonstrator opening three locking mechanisms to get into a box: first, a bolt was pushed out from its moorings, then a hairpin screw was removed from a catch and finally the catch was opened. Birds that observed a demonstration were faster at getting into the box than birds that didn't. However, they didn't do it the same way – both in terms of the order of actions and how they acted on each of the mechanisms. The kea, it seemed, had learned about the properties of the locks, not the actions of the demonstrator.

One solution to control for emulation was to remove the demonstrator from the demonstration, a method known as 'ghost control'. Now, a 'ghost' operated the apparatus (in reality, some clever invisible mechanism devised by the researchers), making the components move without any social stimulus. If observers still solved the problem, they were presumed to have learned how the task worked (i.e. emulation). In contrast, if they failed in this condition but a separate group succeeded with a demonstrator, they were presumed to have learned by imitation.

You'd be forgiven for thinking that two-action tasks with ghost control would have settled the debates around what imitation is and how to demonstrate it. You would be wrong.

The second argument is a bigger criticism and it's been levelled at much of the field. It comes down to novelty. There's a difference between copying which direction to push a lever, for example, and learning to copy a new dance move. The former makes use of actions that are already part of an individual's behavioural repertoire (such as pulling, poking or twisting something), but are used in a novel context, while the latter involves producing movements that are not already encoded in the nervous system. Historically, there's been a bit of a split between those interested in finding laboratory evidence for imitation, whose two-action studies involved the former, and those who were more interested in how innovations (i.e. novel behaviours) could be socially transmitted in a natural group. It's not just a semantic difference: to copy a novel action you must pay attention to what the demonstrator is doing and translate this into what you do, which involves more cognitive processing than producing an existing action from your repertoire. Byrne calls this the 'correspondence problem' and it's something for which I feel an affinity, being *that* person in your dance class who tries to hide at the back so you can't see that I'm moving the wrong part of my body, to the wrong rhythm, sometimes even in the wrong direction.

In an influential publication in 1998, Byrne and psychologist Anne Russon set out a compelling argument for a rethink of imitation. In it, they urged the scientific community to move past asking, 'Is it imitation or not?' and instead focus on the

what and why of the behaviour. They singled out the
two-action task, disputing that any of the studies using it
were actually showing imitation. Instead, Byrne and Russon
proposed that a cognitively simpler mechanism called
'response facilitation' could account for success in these tasks.
It goes like this. If an observer watches another animal receive
a reward using an action that resembles one in its own
repertoire, its corresponding neural 'record' will be primed
and it will be more likely to produce a matching response. It
hinged on the fact that the demonstrated actions already
existed in the animal's repertoire – in essence, as Byrne and
Russon's wrote: 'Novelty will prove to be a cardinal
requirement of imitation.'

Byrne and Russon proposed dividing the category of
imitation into two types: 'action level' and 'programme level'.
They equated the former with impersonation, the traditional
sense of imitation where specific actions are copied. The
latter, they proposed, involves copying not the precise actions
of a demonstrator, but the overall organisational structure of
its behaviour. In doing so, they wanted to bridge the 'wildly
discrepant views' of those (largely fieldworkers) who were
convinced that apes could imitate, and those (largely lab
workers) who were convinced that they could not. Their
own field observations with gorillas and orangutans provided
support. Byrne had documented the highly skilled technique
that mountain gorillas use to process plants such as stinging
nettles, while Russon, working with rehabilitant orangutans
in Indonesia, had observed the animals imitating human
behaviours including siphoning fuel from a drum into a
jerrycan, weeding paths, mixing ingredients for pancakes and
washing dishes. In both cases, imitation of the behaviour
seemed to be at the coarser scale of learning the overall
sequence of behaviours, with the animals using their own
experience to fill in the finer details involved at each stage.

This, believe it or not, is just a snapshot of the theoretical and methodological discussions had by students of social learning over the past century. It is not comprehensive nor definitive – it can't be, given there's still no overall consensus! What it does illustrate is that researchers do not use the term 'imitation' lightly. It's worth bearing that in mind as we get back to monkeys – they were touted as the 'most imitative' of all animals long before definitions were attempted. And surely, with so many tales of monkey copycats, there must be truth to this? The answer is there's some, but a lot more to the contrary. Let's see.

★ ★ ★

Back to the Japanese macaques. Imanishi was inspired to individually identify the animals by the pioneering American primatologist Clarence Ray Carpenter, who established the primate field site of Cayo Santiago. Today, researchers on Cayo Santiago have high-quality behavioural data stretching back for many decades, and none of it provides evidence for imitation. Instead, anecdotal reports strongly suggest the monkeys lack this ability. For example, two brothers discovered how to crack open coconuts by pounding them against rocks, and while several monkeys were recorded gathering around and watching, no other macaque ever learned to do the same.

Unlike Carpenter, Imanishi did not mark the animals: the Kyoto team instead recognised every monkey based on distinguishing features, colour and other bodily characteristics. Koshima was not the only macaque research site, but it was the one that yielded the most well-known findings, including wheat processing, another of Imo's innovations. Wheat was dumped on the beach with potatoes and in the early days the monkeys carefully picked the grains out of the sand. Imo came up with another solution – she dropped handfuls of wheat and sand into the water, and then scooped out the floating grains, enabling her to access more of it faster. Most other monkeys followed suit. For Imanishi and team, this and potato washing were examples of 'preculture'; the first such reports for any non-human animals. They were careful to emphasise that they did not equate the macaques' behaviour with human culture, but the label was and remains contentious, for several reasons. First, while provisioning certainly helped to make the animals more visible for study,* it hugely impacted the macaques' natural way of life, including their preferred habitat (before provisioning, they mainly stuck to the dense forested hills; after, they spent much more time on the beach). The innovations and behaviour transmission simply would not have happened without artificial provisioning. Second, the potato-washing behaviour spread exceptionally slowly: just a handful of individuals acquired it each year, leading some to instead ask what prevented the others from adopting it. Third, the notion that any non-human animal had anything like a culture was (and in certain quarters still is) extremely controversial.

Culture, like tool manufacture, teaching and many other behaviours, was once considered a uniquely human trait. It was thought to provide us with a second form of heredity; i.e. a way that information could be transmitted and inherited non-genetically. Other animals, it was widely assumed, could not be capable of such a thing. Today, there are those who remain committed to this belief. But there are more researchers who are convinced by the growing evidence base for animal cultures. In birds, the presence of regional dialects, or accents,

---

*As well as stimulating research, the ease of habituating macaques to people also stimulated the tourism industry. By 1977, there were 37 monkey parks in Japan with provisioned monkeys, enabling visitors to make their own observations of the animals.

is one strand to a huge area of research, which has found remarkable parallels between the learning of song and human language. Cetaceans (whales, dolphins and porpoises) are another notable group: whale song and a technique known as 'lobtail feeding' provide fascinating evidence for cultural evolution. Further evidence for cultures captures a wide diversity of animal groups, including apes, which we'll come to shortly.

Regarding Imanishi's proposal of preculture in the macaques, Whiten acknowledges the critiques, but concludes that 'overall the studies seemed sound, insofar as the innovations spread through social networks, consistent with social learning'. However, he points out that a 'more compelling' example of social learning in these monkeys is stone handling, which was first witnessed in 1979 in a juvenile member of a Kyoto-based macaque troop. It encompasses numerous forms of stone-based 'play': from simply gathering stones together, to rubbing or cuddling them, to pushing them along the ground, and hitting them against other stones or surfaces. Now reported in several other troops, the 'repertoire' of each troop has expanded over the years, both in number and complexity – combining stones with other objects only emerges after simpler forms of stone handling are established. What this all tells us is that behaviour innovations can spread within a group of Japanese macaques, but the observations to date provide no evidence for imitation. Most likely, the monkeys either independently learned about features of their new, beach environment, or they learned socially through simpler attentional or motivational processes.

For a long time, there was prejudice among primatologists that only great apes and Old World monkeys were capable of intelligent behaviour, because of their closer evolutionary history to us. During the 1980s and 1990s, however, in large part due to the surge of interest in primate social cognition generated by Whiten and Byrne's Machiavellian intelligence hypothesis, primatologists started flocking to the neotropics to study New World monkeys. They were immediately hooked.

Perhaps the most intensely studied New World monkeys are the capuchins, which are found across Central and South America. By the end of the 1990s, they had earned themselves the label, 'apes of the New World', with researchers pointing out the existence of several similar traits. Capuchins are not large monkeys – weighing up to about 4kg – but they are long-lived (up to 55 years in captivity) and have strikingly large brains for their body size. Living in groups of 10–35 individuals, these generalist omnivores forage for all sorts of different foods, including extracting insects and larvae from crevices, and they also hunt small mammals, birds, snakes and reptiles. They are also the masters of monkey tool use, usually to access food, but also in more unusual contexts – female bearded capuchins, for example, sometimes throw stones and branches at the dominant male to get his attention during courtship. Capuchins are exceptionally tolerant of other individuals and, as we heard in the previous chapter, extremely cooperative. This suite of behaviours shows intriguing overlap with those suggested as crucial for the emergence of material culture, making capuchins particularly interesting for social learning research.

As curious animals that are easily trained (think of Ross's monkey 'Marcel' in *Friends* or the monkey in *Outbreak*,[*] among others), capuchins readily participate in captive studies. Elisabetta Visalberghi of the Research National Council in Rome and Dorothy Fragaszy of the University of Georgia in the US, have been leading this research since starting a life-long collaboration in 1984. At this time, scientists and the

---

[*] Fun fact: they were played by the same superstar monkey – a capuchin named Katie. Clearly, she was so good that the directors of Outbreak decided they had to use her even though the storyline concerned the spread of disease from an African (i.e. Old World) monkey.

public alike believed that monkeys were prolific imitators, but as Visalberghi tells me: 'Imitation was defined in so many different ways that it was difficult to discuss this topic among scientists.' Nevertheless, as the pair examined the research more closely, they realised that there was a discrepancy in the popular and scientific literature, something they summed up in a 1990 publication entitled 'Do monkeys ape?' She continues: 'Doree and I highlighted a mismatch: on one side many cultures and languages portrayed monkeys as skilful imitators and on the other there was a lack of evidence that they learned new things by imitation. This was puzzling to us.' They decided to test it, conducting study after study to understand whether and how capuchin monkeys learned from others. In one study, Visalberghi was intrigued to find that even after more than 70 'lessons' from an expert monkey, capuchins never learned the seemingly simple task of inserting a stick into a tube to push out food. Interestingly, she found that the monkeys paid a lot of attention to the demonstration, but they focused on the reward and not the action of inserting the stick. After the lessons, they frequently contacted both the tube and tool, but not the tube opening. In Georgia, Fragaszy's capuchins were doing no better on equivalent tasks, leading the pair to conclude in a 2004 review that capuchin monkeys did not learn by imitation.

While Visalberghi remains convinced of this, she has been conducting further research on the social biases affecting capuchin learning. Experiments with her colleague, Elsa Addessi, have found that although the monkeys don't copy the choice (between two foods) made by others in their group, seeing other individuals eating prompts the observer to start eating no matter what, even if they are already full!

Extensive study has also been made of capuchin nut-cracking in wild and semi-wild populations. This behaviour was first reported by Spanish naturalist Gonzalo

Fernandez de Oviedo, who described it in his Sumario de la
Natural Historia de las Indias (1526):

> Some of these cats* are so astute that many things they see
> men do, they imitate and also do. In particular, there are
> many that when they see how to smash a nut or a pine nut
> with a stone, they do it in the same way and, when leaving
> a stone where the cat can take it, smash all that are given to
> them. They also throw a small stone of the size and weight
> of their strength, as would be thrown by a man.

It was not until 2001 that the first systematic study of
spontaneous nut-cracking behaviour in capuchins was
published, entitled 'Semifree-ranging tufted capuchins (*Cebus
apella*) spontaneously use tools to crack open nuts'.
Nut-cracking is a specialised and skilful behaviour, involving
coordination and careful manipulation of three objects. The
hard palm nut is placed on top of a flat 'anvil' (wooden or
stone) and the animal then uses a 'hammer' stone to break it
open. It's not an easy task – palm nuts are about 20 times
more resistant to cracking than walnuts. The hammer is so
heavy† that a monkey must stand up to lift it (in some cases
even jumping) before bringing it crashing down on to the
nut. Accuracy is crucial, as the hammer could easily crush a
foot. For many years, the assumption was that young monkeys
learned to crack nuts by copying experienced monkeys in
their group. Detailed observations by Visalberghi, Fragaszy

---

*Oviedo referred to monkeys as *Gato monillo*, meaning 'little monkey
cat'.

†In Boa Vista, hammers typically weigh around 1kg, which equates to
30 or 50 per cent of an adult capuchin's weight. Adult females weigh
about 2kg and males 3.5–4.2kg depending on their social status.
Incredibly, the heaviest hammer reported was more than 2.5kg which is
an impressive feat of strength.

and Patricia Izar of the University of São Paulo in Brazil, together with colleagues, have helped to clarify this. Although infants start to handle nuts and stones from an early age and pay keen attention to nut-cracking adults, they do not usually master stone tool use until three and a half years of age. Instead of imitating, they learn by socially biased trial-and-error, as Visalberghi explains: 'By being with expert group members, youngsters have the chance to explore and play with stones, anvils and palm nuts. Adults allow them to gather tiny bits of nut leftovers and these rewarding experiences sustain their efforts at nut cracking.' Years of failed attempts help young capuchins to fine-tune the behaviour, leading to initial successes (often with partially opened palm nuts and less resistant nuts, such as cashews) and further learning. And yet, while a lot of it comes down to individual learning, it's not the whole explanation, since isolated monkeys acquire the behaviour at a much lower rate (if at all). The social environment is still tremendously important to support learning: as the researchers have commented, capuchin monkeys don't learn *from* other monkeys, but *with* them.

The shift in focus from controlled laboratory experiments to the study of wild animals has been seen across the field. It reflects a broader change in research questions, with less of a focus on whether animals can imitate, and more on how and why they integrate a range of social learning mechanisms into their natural lives.

Erica van de Waal, professor at the University of Lausanne, Switzerland, has been at the forefront of this shift since she began studying wild vervet monkeys for her master's project in 2005. Some 15 years later, she's still studying vervets – now as director of the Inkawu Vervet Project (IVP) in South Africa, which she established in 2010. Van de Waal's aim is to bring lab cognition methodology to the wild and vervet monkeys are an ideal study system. They're common, opportunistic monkeys that, unlike many primate species, spend a lot of time on the ground and habituate easily to the presence of human observers. They're also social and, like capuchins, readily tolerate the presence of other individuals as they forage and feed. The IVP now includes around 200 individually recognised wild monkeys belonging to six groups, offering a wealth of research opportunities.

Van de Waal was initially interested in whether the vervets could copy each other and what process accounted for their behaviour. In her first experiment she introduced a puzzle box to each of the six groups and let whichever monkey was monopolising the box be her demonstrator. It happened to be the dominant female in three groups and the dominant male in the other three. She found that the other monkeys learned from the demonstrations, but only when the demonstrator was the dominant female, an intriguing finding that changed van de Waal's research focus: 'It directly brought me to the, I think, much more exciting topic of social learning biases – when do you learn socially and from whom?' Further studies have confirmed the female bias, but there's also flexibility: when pay-offs were manipulated so the male demonstrator accessed five times more food than the female, male observers switched to copy him while females continued to follow the female.

A key advantage of van de Waal's study system is that she can evaluate the spread of behaviour in wild groups, meaning that she can study the formation of traditions in as natural a way as possible. Together with Whiten, she built another puzzle box to investigate the transmission of feeding behaviours. The apparatus had a door that could be lifted or slid to either side and a demonstrator was trained to open it in one of these ways. The demonstrator and several puzzle boxes were then placed into each of three groups, with the individual monkeys free to watch and interact with the apparatus. Van de Waal was surprised by the results. 'We could really see that they copied the opening technique,' she says. 'Whether it is really imitation or not depends on your definition ... I prefer "bodily matching" because it doesn't imply that you're imitating a novel action. They didn't have to learn a new movement, they just had to match a movement that they do naturally to something else.' In contrast, when presented with a task involving a sequence of different actions, the monkeys only learned which part of the apparatus to touch, not what to do. This suggests that the sequence was too different to anything in their existing repertoire to be matched.

In a now-classic study, van de Waal and Whiten focused on the transmission of food preferences rather than actions. Four vervet groups were each provided with two troughs of colourful corn, one dyed pink and the other blue. The corn in one of the troughs had been soaked in bitter aloe leaves and so tasted disgusting: for two of the groups this was the pink corn and for the other two it was the blue. The monkeys could freely feed from both troughs and learn which corn tasted bad, and the troughs were then removed for four to six months. Upon reintroduction, none of the coloured corn was distasteful, but the monkeys continued avoiding the colour that had previously tasted bad. Of most interest was the behaviour of new monkeys to each group, which included babies that had been born in the corn-free period and males that emigrated to a neighbouring group. The results were astonishing. All 27 of the infants that had been born in the corn-free interval ate the colour that their mothers ate, ignoring the full trough of perfectly edible corn right next to it. That was pretty much expected. More excitingly, of the 15 males that dispersed to other groups (fortuitously, all joined a group for which the other colour of corn was the norm), all but one of them quickly switched their preference to match that of their new group. 'This was the big "wow" moment for me,' says van de Waal. 'I saw first one conforming, then two days later in another group another one and then a week later another.' It was a fascinating outcome – the males had an entire trough of their preferred colour available; they could eat as much as they wanted, but seeing their new group fighting over access to the other colour caused them to switch their preference. As van de Waal sums up: 'They put a higher weight on what they learned from others than what they had learned themselves.' They were conforming to social norms, something that many people had assumed to be a human trait or, possibly, human and chimpanzee. Its presence in vervet monkeys surprised many in the wider research community.

A key ongoing research theme is understanding more about why animals learn from each other, van de Waal explains. Humans learn all the time about social norms and how to fit in, yet there's a perception that animals only learn socially to gain novel information. Intriguing research with other monkeys suggests this may not be the case.

Susan Perry, professor of anthropology at UCLA, founded the Lomas Barbudal Monkey Project in Costa Rica in 1990, and has been studying the formation, maintenance and disappearance of traditions among white-faced capuchins ever since. Early on, Perry realised that most research on animal traditions had focused on foraging techniques, whereas in human societies, social conventions arguably occupy more of our cultural repertoire. We unconsciously mirror the mannerisms, gestures and language of people that we are close to. Termed the 'chameleon effect', it has been shown to promote social bonding, with those being imitated reporting feelings of increased social rapport and empathy. Perry's research provides fascinating evidence that the same may happen in capuchin groups.

In a major cross-site collaboration with a team from three other study sites on the island, Perry and nine other researchers analysed more than 19,000 hours of data from 13 years of research with 13 social groups. They found consistent differences between groups in three classes of behaviour:

hand-sniffing, sucking of body parts and games. Usually the behaviours were reciprocal. Hand-sniffing, for example, involved two monkeys sitting together, each one holding the other's hand against its nose or inserting its fingers inside, and breathing deeply with a trancelike expression on their faces. More recently, Perry has started to document the spread of a new tradition, 'eyeball-poking', in which one individual inserts its finger between the eyelid and lower eyeball of its 'friend', sometimes up to the first knuckle in depth. The common theme for all the behaviours is that they involve a level of risk, or at least discomfort, to both participants. The current hypothesis for such bizarre traditions is that they test the strength of social bonds, which may be important in knowing which individuals to trust. Importantly, the behaviours were not universally recorded across all capuchin populations, nor were they permanent. Instead, they popped up in some groups, becoming established as traditions for a variable amount of time – up to 10 years for some – and then disappeared. The fragility of these traditions is likely down to the disproportionate impacts of a few knowledgeable or enthusiastic individuals – when those individuals disappeared, the behaviour often did too. Perry was initially sceptical that imitation played any role in the acquisition of these behaviours, but has not dismissed the possibility – it may be that for monkeys, this form of learning is more relevant to the building and maintaining of relationships, rather than the learning of new technical skills.

Studies of captive capuchins and pigtailed macaques provide additional evidence for the idea that imitation has a social function; what's more, it's not restricted to members of the same species. In these experiments, conducted partly at Visalberghi's lab in Rome, each monkey was provided with a small object that it could manipulate (a wooden cube for the macaques and a ball for the capuchins) and two experimenters a short distance away were also provided with the same. One experimenter imitated all of the monkey's actions (e.g. hitting the object on a surface, biting it, etc.) while the other manipulated the object at the same time as the monkey, but in a random way. Macaques and capuchins both spent longer looking at the imitator than the non-imitator. What's more, in the capuchin study, monkeys that had previously learned they could exchange tokens with human experimenters for rewards were far more likely to exchange with the imitator, even though both experimenters provided the same reward. It seems that the monkeys found something compelling about the person who copied them, suggesting that, just as in humans, imitation may promote social bonding and cooperation. As Visalberghi comments: 'This "tuning" might be advantageous for animal species in which the social life is very important, as is the case for primates (humans included).'

How do monkeys know that they're being copied? The final part of the story dates to the early 1990s, when a team of researchers led by Giacomo Rizzolatti at Parma University in Italy stumbled upon something unusual in their neurophysiological studies of macaques. The team had implanted tiny electrodes into the animals' brains to record how individual neurons were firing under different scenarios; their aim was to work out how the brain coordinated and controlled the fine movements of muscles in the hand. The expectation was that specific neurons would fire under different circumstances; when the animal reached for something, when it grasped an object, and so on, and this was indeed what they recorded, but it wasn't the full story. Remarkably, the team observed that some of the same neurons also fired when the monkey watched an experimenter reaching for and grasping food, an accidental finding that came about when one of the team happened to eat something in view of the monkeys one day. The results were published in 1992 and in 1996 the team dubbed their discovery 'mirror neurons', suggesting that they may provide the neurological basis for the monkey translating what it saw into what it did (i.e. a neural explanation for the mechanism of response facilitation suggested by Byrne and Russon). Mirror neurons – by firing both when an individual performs an action and when it observes that action in another – are thought to provide a powerful way for individuals to automatically understand others' intentions and emotions. This unplanned, serendipitous observation kickstarted several exciting new avenues of research, predominantly in humans, including the study of imitation as well as theory of mind, mirror self-recognition, autism and the evolution of language.

Fact or fiction?

Monkey see, monkey do. It's a term that has persisted for centuries, one that has well and truly sunk into our cultural consciousness. Monkeys are undoubtedly rapid learners and they show abundant evidence of learning from each other. Diffusion experiments show that behaviours can be transmitted through a group, providing experimental support for the spread of Imo's and the other monkeys' innovations in Japan. The overall impression from all of the studies is that monkeys are unable to learn by imitation, although they do recognise when they are being imitated. This means that Aesop was wrong, but it indicates something interesting: that imitation is not necessary for the formation of local traditions, in both the foraging and social spheres.

If not the monkey, then what is the 'most imitative' animal? Let's first address the elephant in the room that is vocal imitation in songbirds and parrots. Undoubtedly, some birds show remarkable precision in their vocal matching – something that has shed light on our own acquisition of language. To do this field justice would take another chapter, but it's also a different kind of imitation. Since Aesop's monkeys were imitating behaviours, I'll stay focused on behavioural imitation here – and for that the great apes make much better candidates.

The scientific study of ape imitation dates back more than a hundred years. One of the first published accounts was American psychologist Lightner Witmer's 1909 article, 'A monkey with a mind', which outlined Witmer's experience with a roller-skating circus chimpanzee named Peter. Witmer set up various tests for Peter, including the 'cigarette and match test', in which we are told that 'Peter lights and smokes his cigarette as intelligently as a man'.* One test focused specifically on imitation. Witmer picked up a piece of chalk and drew a 'W' on a blackboard in front of Peter, who 'watched the operation intently, then with the chalk in his hand … quickly made the four movements and drew a fairly perfect letter beneath the "W" which I had traced'. Witmer concluded that Peter's imitative ability was equal or superior to that of many children his age and predicted that further insights could be made if an animal was hand-reared from an early age as a human child.

Psychologists Keith and Catherine Hayes of the Yerkes Laboratory of Primate Research in Orange Park, Florida, did just this. In 1947, they adopted an infant chimpanzee called Viki and brought her up in their home, even dressing her in infant clothes. The Hayes pioneered an approach to studying imitation now termed 'do as I do' – they trained the young ape to copy a series of arbitrary actions on the command 'do this' and reported that after the first 11 demonstrations she seemed to get the concept, subsequently generalising the rule of copying to the new demonstrated actions. In a progress

---

* Clearly, research was quite different in those days.

report when Viki was three years old, the Hayes' commented on her propensity for imitation. 'Just as a human child copies its parents' routine chores, so Viki dusts; washes dishes; sharpens pencils; saws, hammers and sandpapers furniture; paints woodwork; and presses photographs in books.' Viki contracted viral encephalitis in 1954 and died shortly before her seventh birthday. Had she survived to the average age of a chimpanzee (40–45 years in the wild) then we would likely have seen evidence for a greater diversity and complexity in her behaviour.

At the core of these early studies was a desire to learn how much like a human a chimpanzee could become. Key questions were whether the animals could develop human language,* and how their development overlapped with that of human children if raised under the same conditions. Some exceptional insights were made but, as we've seen in other fields, the study of animal cognition has changed considerably, and today most researchers consider the natural ecology and evolutionary history of their study species when formulating any questions about their behaviour. From this perspective, one controversial question is how useful imitation really is to a wild animal?

'Simply imitating is a really poor strategy,' says Byrne. 'And that fits in with Aesop's view – he saw monkeys as things that did nothing but imitate. Real monkeys are curious and explorative, and in social circumstances they benefit from stimulus enhancement and social facilitation in learning new skills – whether or not they can or ever do imitate.' For

---

*The ability for spoken language was a key driver for these early studies, but neglected important facts about chimpanzee natural behaviour. Chimpanzees are highly communicative animals, but they tend to vocalise only when excited; for example, whooping when being challenged within the group. Calm chimpanzees are usually quiet, communicating with their groupmates through grooming and gestures.

Byrne, Russon and others, the programme-level imitation
seen in apes could be useful for acquiring complex, novel
sequences of behaviour, in the same way that an infant learns
to tie her shoelaces. Action-level imitation isn't a useful way
for an animal to learn about the world, unless it's in a really
dangerous situation and needs to rapidly decide how to act:
doing what everyone else is doing is, after all, better than
dying. But, Byrne counters, it's not smart because it won't
lead to innovations. Trial-and-error learning should be more
useful, because the animal acquires a behaviour that is
specifically adapted to the problems it faces and its own
competencies. 'Imagine a little gorilla imitating his mum,' he
says, referring to nettle processing. 'Her hands are almost as
big as he is. If he tried to impersonate her actions precisely,
he'd fail completely and probably get stung! But if he watched
and understood the sequence of functional steps, he might
start to achieve each step in his own way, imitating the
programme not the movements.'

The function of action level imitation (impersonation),
Byrne and Russon proposed, may be in social bonding, as we
saw with Perry's capuchin research. Whiten agrees with this,
but takes a different view about the value of imitation: 'What
imitation contributes is the value of copying actions, especially
complex ones.' He explains that a 'copy first, refine later'
approach may be particularly important for humans, because
'the world of actions and artefacts is so complex and often
causally opaque, it's a good first strategy to copy what a
competent adult does, then experiment later, discovering that
some things you copied were not essential'. The true power
of imitation over other forms of social learning is that repeated
imitative copying can give rise to stable traditions. This is
where social learning ties in with culture – just as the Kyoto
team had suggested more than 60 years ago.

The impact of Imanishi and his students was not restricted
to Japanese macaques. In 1965, Jun'ichiro Itani established

a research population of chimpanzees at Kasoge in Tanzania. These were the eastern subspecies, like at Gombe (the populations were once contiguous, but deforestation had formed a geographic barrier), so offered an ideal system in which to study the presence of local traditions. In 1963, Jane Goodall, while still a PhD student, asserted that the tool use and manufacture demonstrated by the chimpanzees at Gombe revealed a primitive culture, a position that she emphasised a decade later by presenting data on different behaviours from different study sites. Interest was growing, although it led to much heated debate. Bill McGrew, honorary professor at the University of St Andrews, while making observations of the Kasoge chimps with PhD student Caroline Tutin in 1975, documented something called the 'grooming–hand–clasp'. Sitting facing each other, both animals lifted either their right or left arms, clasped the other's hand or wrist and proceeded to groom the exposed armpit. The behaviour had never been described from the many thousands of hours spent observing the Gombe chimpanzees, so in their 1978 report McGrew and Tutin suggested it should qualify as a 'social custom'. Even though they intentionally avoided the label of culture, McGrew has recalled how their publication was immediately 'broadsided' by anthropologists, something that marked the beginnings of what he has termed the 'chimpanzee culture wars'.

Although the observations of Goodall, McGrew and others suggested the presence of local traditions in different chimpanzee populations, they could only provide part of the story. To see the overall picture, the different research teams needed to join forces and pool their data so, led by Whiten, they did just that. In 1999, the collaboration published a pivotal analysis from the seven longest-running field sites, comprising a monumental 151 total years of chimpanzee observations. It revealed 39 different behaviour patterns that were common at some sites but not at others, including food processing, tool use and grooming techniques, and courtship displays. The western group of one subspecies, for example, was proficient at cracking open nuts with rocks or pieces of wood; however, just across the wide Sassandra-N'Zo river, the eastern population of the same subspecies did not crack nuts. For the team, this was evidence that nut-cracking behaviour in the western population resulted from social transmission, as opposed to genetic or environmental influences. The same conclusion was reached across the 39 behaviour patterns, providing convincing evidence for chimpanzee cultures. Subsequently, studies have revealed a similar story for orangutans and gorillas. Together with cetaceans, birds, fish, rats and others, evidence is mounting for cultures in a range of other species. Yet, no other non-human animal comes close to the chimpanzee for the number and diversity of socially transmitted behaviours; for the fact that each group of chimpanzees shows evidence of its own culture.[*]

The observational data from wild groups were convincing, but what they couldn't reveal was how the behaviours were transmitted. For this, an experimental approach is necessary, similar to that developed by van de Waal with vervet monkeys. A good example comes from one of Whiten's experiments with collaborators Victoria Horner and Frans de Waal. Two female chimpanzees ('Georgia' and 'Ericka') were each trained to retrieve food from a 'pan pipes' apparatus using a different technique. Once proficient, they were released back to their respective groups and the apparatus set up in their

---

[*]Which is why recent studies suggesting that human presence and habitat fragmentation is leading to the erosion of chimpanzee cultures should make us all stop and think. Behaviours that have taken generations to become established, that make each group of chimpanzees special, could simply vanish due to our activity.

enclosures. There was no ambiguity in the results: Georgia's technique spread within her group, and Ericka's technique spread within hers. A third group, which never saw any demonstration, didn't figure out how to get the food. Interestingly, in both groups some individuals spontaneously discovered the other technique, but they quickly reverted to the technique of their group. Just like the young male vervet monkeys, social conformity trumped individually learned information for the chimps.

The next question was how the behaviour spread – were the chimpanzees learning by imitation? One hundred years ago, the answer would likely have been a definitive 'yes' but, as we've seen, the consensus on imitation has waxed and waned in more recent years. This is no bad thing: there is general agreement that the debates and critiques of earlier work resulted in increasingly rigorous experimental design and interpretation, although there's clearly some way to go for a consensus. Nonetheless, some ingenious experiments have been devised to try and get to the heart of what it means 'to ape'.

Another of Horner and Whiten's studies investigated imitation and emulation learning with chimpanzees and pre-school children. Taking advantage of the chimpanzees' natural use of tools to 'fish' for termites, they developed an artificial termite mound – in reality, an opaque Perspex box with covered holes in the top and side – and Horner, a familiar play partner, acted as the demonstrator. She used a stick to remove a cover over the top hole, then stabbed the stick firmly into it several times. She then removed the cover on the side hole, inserted the stick and retrieved a food treat. The mound was then baited with food again and the chimpanzee could take its turn. Horner found that the animals did copy her sequence of actions, showing they could imitate. But a separate group did something very different. They had received identical demonstrations, but with a transparent box, meaning that they could see there was a barrier between the food and the hole in the top; i.e. they could see that the first behaviour (stabbing the stick into the top hole) was causally irrelevant. The transparent box revealed that the most efficient way to retrieve the food was to simply open the cover from the side hole and fish it out. Horner and Whiten predicted that if the chimpanzees were 'rational imitators', they shouldn't imitate the full sequence of events with the transparent box and this is exactly what they found: the animals took a no-nonsense approach and did only what was necessary to get the food. In this case, emulation was the more efficient strategy. A more recent study found the same with bonobos: they imitated when no other information was available, but never copied any action that was visibly causally irrelevant. These studies show that chimpanzees and bonobos can be flexible in the way that they learn, taking the rational approach when the information available to them allows it.

And what about the children? As Whiten tells me: 'We did that experiment rather expecting that the more intelligent species, humans, would discriminate between the transparent and opaque box conditions, and the interesting question was whether apes would too.' In fact, in the transparent condition, children continued to copy the demonstrator even though they could see that poking the stick into the top hole had no effect on the position of the food. By copying this clearly irrelevant action, children were doing something that was subsequently termed 'over-imitation'. It set the wheels in motion for a new area of interest, particularly among developmental psychologists: since Horner and Whiten's study was published in 2005, more than 50 studies have been conducted on the topic, mostly in children and with a few confirming the original results with apes.

On the face of it, it seems paradoxical that humans should be the ones that blindly copy others. It is, after all, what Aesop's monkeys did. Over-imitation seems foolish and yet it seems to be a key component of being human. Controversy abounds on the reason why, but most researchers agree that what appears at first sight to be a counterintuitive trait is fundamental to the extreme cultural nature of our species. Certainly, our propensity for imitating seems to set us apart from the rest of the animal world. And since in everyday life most people's intentional actions have a function, over-imitating might not be so irrational after all. When we're unsure how something works, then copying all the actions we see may be a good rule of thumb, with subsequent refinements being made as we learn what is relevant and what is not.

Faithful transmission of knowledge or skills means they can be preserved or improved over generations. Our civilisations stand on the shoulders of giants, building on previous enhancements to continue to advance. This is called cumulative cultural change and it has been likened to a ratchet by Tomasello, who proposed that cultural traditions or artifacts accumulate modifications over time. Undoubtedly, no other animal demonstrates anything like a human level of cumulative culture.

Conformity is also hugely important in human society. As much as we might strive to stand out from the crowd and demonstrate our individuality, a fundamental part of being human is behaving in ways that don't get us kicked out of our social group. The importance of group membership is highlighted by findings that rejection activates brain circuits related to physical pain. We need to belong and conforming to social norms is one tool that can help us achieve that. That this is clearly also important in ape and monkey social life shows that we are not unique in that respect.

We previously saw that capuchins and macaques recognise when they are being imitated, and apes do too – but they also do something more cognitively complex. In one study at Leipzig, all four species of ape additionally responded to the imitator by producing what are known as 'testing' behaviours. These intentionally produced actions are characterised by being sudden or unusual in their form or timing, and it's thought that they are produced to test the contingency between the two interacting individuals. Children, for example, might show a deliberately bizarre behaviour towards someone who is imitating them to see how far they will go to copy them. Their production by apes and humans, but not monkeys, lends additional weight to the view that although monkeys possess mirror neurons and the capacity for behaviour matching, they have a much simpler capacity for imitation.

Imitative interactions were also documented in a study at Furuvik Zoo in Sweden, which looked at how visitors interacted with the chimpanzees and vice versa. Researchers found that about 10 per cent of all actions made towards the other species were imitative. What's more, cross-species interactions where the individuals imitated each other lasted longer than those that didn't, indicating that like the macaques and capuchins, there was something compelling about the behaviour. It's a fascinating set of observations and one which saw the team being awarded the 2018 Ig Nobel Prize in Anthropology 'for collecting evidence, in a zoo, that chimpanzees imitate humans about as often, and about as well, as humans imitate chimpanzees'.

★ ★ ★

The study of animal social learning has been long and convoluted, and controversial questions remain around what and why animals learn from one another. Underpinning all of this is a fascination with the concept of imitation, although interpretations of it have oscillated between it being a 'cheap monkey trick' to the holy grail of animal learning. Knowledge in the field has greatly increased, and with that has come a change in focus – no longer are researchers on a quest to find out simply if their study species imitates; more nuanced questions are being asked about who they learn from, when and why. Visalberghi tells me that one of the problems with focusing on evidence for imitation is that, 'unfortunately, laymen are mostly exposed to what attracts the media's attention, to what can be summarised in a few sexy words'. I hope, from this chapter, it's clear that imitation is far from simple or common in the animal world. The reality is that a lot of behaviour labelled as 'imitation' in the media reflects a complex interplay of social influence, innate bias and individual learning, which is far less easy to neatly summarise!

Humans are the real masters of imitation – we are the only animals for which copying the actions of others is so fundamental to our cultural evolution that it can take precedence over being rational, and we've been dubbed 'Homo imitans'* for this reason. Ape or whale cultures may be complex compared with other animals, but they pale in comparison with our own. That's because while animals may have tens of known behaviour traditions, ours are limitless. We can transmit not just specific behaviours through populations, but entire fields of knowledge too. But we can't take the monkey's place, so our attention turns to the apes and, given the amount of research that has been done with them, it's got to be the chimpanzee. The trouble for our fable

---

*First coined by developmental psychologist Andrew Meltzoff in 1988.

is that although chimps can imitate, the evidence shows that they don't do so blindly if they can see a more rational alternative. The chimpanzee might not have gotten into the sort of scrapes that were described for the monkey, because it would have already worked out the best thing to do.

CHAPTER EIGHT

The Ants and the Grasshopper

The ants were spending a fine winter's day drying grain collected in the summertime. A Grasshopper, perishing with famine, passed by and earnestly begged for a little food. The ants inquired of him, 'Why did you not treasure up food during the summer?' He replied, 'I had not leisure enough. I passed the days in singing.' They then said in derision: 'If you were foolish enough to sing all the summer, you must dance supperless to bed in the winter.'

During the summer of 2005, I set up some experiments to investigate whether the crows would use a little tool to retrieve a longer tool when necessary, as described in Chapter 1. Essentially, I was asking if the birds could plan out their behaviour. All the trials took place in a small, brightly lit testing room, on a table that was flush against a large one-way window. On the other side of the window was an equally small experimenter's room and this is where I sat, in the dark, huddled over my laptop or behind a video camera, watching the birds for many hours at a time.

A little hatch connected each of the two main aviaries to the testing room. We controlled its opening from inside the experimenter's room, but once the bird was in the testing room it could leave whenever it wanted. Usually, this was after it had solved the task and eaten the reward, but it could also be if the bird became startled by a noise outside or frustrated by being unable to solve the task. In a couple of these trials, a bird named Pierre left the testing room after trying and failing to retrieve the reward with the short tool, and before I had locked the gate and set up for the next trial he snuck back in with a stick from his aviary, which was long enough for him to go straight for the food.

Part of the excitement of animal behaviour research is that you can be surprised like this. No matter how well you think you've planned an experiment, trying to ensure that the animal can only pick up *this* object or press *that* lever, animals can and will find a way to thwart you. Pierre shouldn't have been able to retrieve the food the way he did – it wasn't one of my predefined outcomes and it complicated my analysis. But it showed me, albeit anecdotally, that his ability to solve problems was more flexible than I had imagined, which made the extra statistical complexity forgivable. Pierre was a good experimental bird: he was used to coming into the testing room and participating in our experiments, and he got used to new apparatus and procedures quickly. But, as well as sometimes bringing his own tools into the testing room, Pierre developed a habit of removing them as well – and, unfortunately, he was much quicker than me. Sometimes, I could see what he was about to do and would dash into his home aviary quickly enough to startle him into dropping the tool. Other times, I was too slow and by the time I got inside, there was Pierre, no tool in sight, hopping between perches and squawking at me. After fruitless searching I would then have to go and make a new tool, which was a pain because it interrupted my experimental time and because, let's face it, no one likes being outsmarted by a crow. I was left simultaneously cursing Pierre and scratching my head about where the darned tools could be.

The mystery of the missing tools was solved a few weeks later, when we gave the aviaries their routine scrub down. There, under a corner of tarpaulin that was usually buried under three inches of bedding material, I found a little hoard of treasure. Pierre's treasure. A few less-than-fresh chunks of his favourite food (pig's heart), a couple of Lego blocks and three of my experimental tools. It was clearly not an accidental collection of objects that had somehow washed up together in that corner of the room. Pierre had put them there, intentionally, a realisation that left me amazed. Was he planning for the future by saving his favourite tools?

Aesop's ants are depicted as responsible, future-thinking animals, 'squirreling' away supplies to last them over winter. The grasshopper, on the other hand, was concerned only with present desires: frolicking in the sun all summer, it spared no thought for future needs. My sequential tool-use experiment showed that the crows could plan out appropriate tool sequences when required; however, this was still confined to the present. The hoarded tools, on the other hand, prompted the tantalising thought that Pierre may be thinking further ahead. Of course, an anecdote like this doesn't tell us anything about why Pierre stashed the tools, but fortunately future-planning is a hot topic in animal behaviour research. So, what's the evidence – are Aesop's ants grounded in fact or fiction?

★ ★ ★

Approximately 90 years before me, German psychologist
Wolfgang Köhler was wondering similar things about a group
of captive chimpanzees in his research station on Tenerife.
Although, as we heard earlier, Köhler was confident enough to
claim the presence of insightful problem-solving in some
subjects, he acknowledged an important limitation to the
chimps' behaviour, writing: 'The time in which the chimpanzee
lives is limited in past and future.' He described one example of
potential planning in an ape called Sultan, who sharpened the
end of a stick so that it could just fit inside a tube and make a
'double tool', long enough to reach a piece of fruit. Köhler
reflected that although this suggested an ability to plan, Sultan's
behaviour was not sufficient to claim future thought:

> To be sure, there is, in the example given, the incentive
> of the visible reward, and, all through his labour, he could
> glance from time to time at the fruit. Anyone seeing an ape
> making preparations for an anticipated future experiment,
> the conditions for which are not at the time in sight, would
> be witness of a far higher achievement in the direction
> under discussion.

Köhler's ideas were re-popularised in the 1970s by Norbert
Bischof and Doris Bischof-Köhler, whose influential work
proposed that non-human animals are 'stuck in time'.
According to their theory, which remains highly influential,
other animals' behaviour is driven entirely by their current
feelings of hunger, thirst, sexual arousal and so on. We, in
contrast, frequently do things that are not related to our
current motivational state: we put money into pensions,
book holidays months ahead of time and stock up the
freezer with foods we have no urge to eat at the time of
buying. We do this because we are able to anticipate future
needs: a stable income many years later, a holiday next

spring, a meal for the friends we've invited for dinner. Supporters of the Bischof-Köhler hypothesis, such as psychologist Thomas Suddendorff, have refined the definition and the latest version of the hypothesis states that 'only humans can flexibly anticipate their own future mental states of need and act now to secure them'. Only we can distract ourselves from present needs to dwell on past memories or fret about the future. Animals, according to this, are the masters of mindfulness, in that they are well and truly in the present.

Is there really such a clear dividing line between humans and other animals, and doesn't this ignore something really obvious? As I write, a squirrel is busying itself out in my garden, nose twitching as it scurries across the lawn, burying and reburying nuts and seeds that it will return to dig up in the winter. Lots of animals do something similar, including pika, jays, moles and many species of bird. As summer draws to an end, vast numbers of birds fly south to warmer climes, and many bears, bats, snakes and others simply shut down and hibernate over winter. When the birds return in spring, they start to construct nests before they have even mated, let alone before any eggs or chicks are present. Surely all of these examples show future planning: they are doing things that enable them to cope better with future events, rather than satisfying a current state of hunger, thirst or sexual arousal. The Bischof-Köhler hypothesis states that this cannot be the case. And being clear on why these examples do not, in fact, reflect mental time travel, is crucial for our fable.

A good way to think of it is that all planning behaviour concerns the future, but not all future-oriented behaviour involves planning. It's a subtle distinction and of course it lies not in *what* the animals are doing, but *how*. That squirrel, making a nuisance of itself by burying acorns in one of my pots, is demonstrating something called anticipatory behaviour. This is a trait that has evolved in members of the same species to allow individuals to respond to significant, long-term regularities, such as the changing seasons. All squirrels store food over winter because all possess the same genetic architecture that enables them to detect the changing seasons, react to changes in day length and switch to burying food. From an evolutionary perspective, it makes sense that individuals of a species that could react to predictable environmental changes would have been more successful than those that couldn't, leading to evolution of that behaviour in the population and the species. Migratory birds have evolved sensitive mechanisms for detecting the changing seasons – such as decreasing daylight hours – and when they detect these environmental cues they are triggered to show appropriate 'planning' behaviour. For example, before birds set off on migration, they 'carbo-load' by gorging themselves on berries and other high-calorie foods to ensure they have adequate energy supplies to last the journey. Bears do something similar: the black bear starts guzzling berries and other energy-rich foods in late summer, and can gain a staggering 13kg every week. The grizzly bears of Katmai National Park in Alaska have become famous for their extraordinary pre-hibernation weight gain, since the launch of 'Fat Bear Week' in 2015. These bears binge on the summertime abundance of salmon, eating a year's worth of food in six months to prepare for losing a third of their body weight during hibernation. During the months of deep sleep, their thick layers of fat and dense fur help to keep body temperature stable, and the gradual breakdown of fat provides water and up to 4,000 calories per day.

Impressive, yes. But cognitively, there is a huge difference between this – let's call it reactive – behaviour and proactive future planning, where the animal mentally projects itself to a point in the future, predicts what might be going on at that point and then uses this prediction to guide what they do in the present. Calling migration or hibernation examples of future planning means assuming that at the end of every summer every year the swallow or bear thinks about the winter, imagines the cold temperatures and lack of food and makes a decision to fly thousands of miles to Africa for a few months, or hole up in a burrow and ride it out. If this doesn't seem too far-fetched it's probably because this is exactly why some people decide to book a holiday to the winter sun. But those sorts of generalisations aren't very scientific and in fact there is good evidence that this is not what migrating animals do. Swallows that are just a few months old and have never experienced a harsh winter still migrate to Africa at the end of their first summer, as do the fourth-generation monarch butterflies, who fly to Mexico just weeks after hatching. How can they be anticipating something they have never experienced?

In our captive colony of New Caledonian crows, we had to house two males together because there weren't enough females for all the birds to be in breeding pairs. Nonetheless, come the spring our boys showed signs of nest-building. It wasn't a skilled, carefully constructed affair, but a rather crude assemblage of twigs balanced on a fork between some branches. Yet it showed that even though these birds were not in a position to mate, they still possessed the instinctive drive to prepare for the breeding season.

The food hoarders, migratory birds and hibernating bears appear to show future planning, because what they're doing allows them to deal more effectively with the future. But not only is there no evidence that the animals are aware of their future needs, the fact that most of them have never experienced the future state they are apparently planning for demonstrates the opposite. So far, this is all consistent with the Bischof-Köhler theory. How does it match up with our ant and grasshopper characters?

★ ★ ★

I must confess that before I started writing this chapter, I knew very little about grasshoppers or their relatives. I certainly had no idea that there are more than 8,000 grasshopper species worldwide, nor that some species live in rainforest canopies and others can swim underwater! They are insects, of course, belonging to the evolutionary order called *Orthoptera*, which also contains locusts, crickets, katydids and the New Zealand endemics, the wētā. With its origins stretching back to the Carboniferous period, approximately 350–300 million years ago, this is one of the most ancient insect groups.

The grasshopper lifecycle follows a reasonably similar pattern, regardless of the species. The male inseminates the female, transferring a little packet of sperm called a spermatophore. The female deposits each fertilised egg 1–2 inches under the ground, using a specialised 'prong-like' adaptation on her insect bottom called an ovipositor. The eggs are grouped together, and the female covers them with a sticky substance that hardens to form a protective pod. Depending on the species, the pod can house anywhere from 15 to 150 eggs, and each female may lay up to 25 pods. That adds up to a lot of eggs, which overwinter and hatch out in May, emerging as adults by June. Initially wingless, they go through a series of moults before they reach full maturity and full adults can survive until early December.

None of this seems to immediately contradict Aesop's portrayal of the grasshopper. Yet in fact, here is the first flaw in Aesop's fable – come the winter, the grasshopper was going to die anyway. Of course, we don't know which species Aesop was referring to, but grasshoppers would have been as commonly seen (and heard) in his time as they are today. Aesop would surely have heard the 'song' of the grasshopper – made by vigorously rubbing its hind legs (which are covered in tiny 'pegs') over toughened veins in its wings – during the summer months.

Searching for evidence about grasshopper planning is more problematic, because although there have been some tests of learning ability, they're clearly not model organisms for behaviour research. That may itself prove the point!

And yet, there is one aspect of grasshopper biology that hints at an ability to plan. This is the remarkable Jekyll-and-Hyde transformation undergone by a dozen or so species when they transform into locusts. Grasshoppers are normally solitary creatures, preferring their own company for their main business of gnawing on vegetation. But one group of short-horned grasshoppers undergoes a transformation of epic proportions when conditions are right; specifically, when a period of rainfall is followed by a drought, condensing individuals into any areas where food is still available. When density exceeds a certain threshold, the grasshoppers change their behaviour and actively seek out other individuals, dispersing over huge distances as a swarm. Not only that, they become stronger, darker and more mobile, which is why, before the 1920s, the two phases of a desert locust life were assumed to be different species. A swarm can cover several hundred square kilometres and wreak phenomenal devastation; crop failure can be a national disaster. 'Biblical' is a term that is often overused, but in this case is spot on, given how this particular phenomenon was one of the 10 plagues inflicted upon the Egyptian Pharaoh in the bible. Indeed, in Arabic the word 'locust' translates as 'teeth of the wind'. It's an incredible biological phenomenon, but it doesn't fit our definition of future planning – changing into locusts is not the result of each grasshopper thinking about the pattern of rainfall and drought, and anticipating future needs; rather, it is an evolved adaptation that increases the survival of grasshoppers under extreme conditions.

Locusts can learn to choose specific food types and also remember their choice over a period of days. We can therefore say that grasshoppers are capable of learning and forming short-term memories, but there's no evidence that they can or cannot think ahead. They hatch, sing through the summer to find a mate, the female lays her eggs and when winter comes, they die. And that means, so far, Aesop was right.

★ ★ ★

Ant behaviour, on the other hand, has been the subject of a much longer and more intense research effort. The first book on ants was Jeremias Wilde's *De Formica*; 15 brief chapters on numerous aspects of ant lives, published in Latin in 1615. People started to make scientific observations of ants in the nineteenth century, but they have featured in scientific and cultural texts that date much further back. Myrmecology is the rather evocative-sounding name given to the scientific study of ants, from the Greek words *myrmex* (meaning 'ant') and *logos* (meaning 'study'), and in fact the ancient Greeks held ants in quite high regard. In Homer's *Iliad* (800 BC), for example, the legendary warrior Achilles commanded the 'myrmidons' (or 'ant people'), who were superior fighters characterised by bravery, skill and loyalty to their leaders.

The more than 12,000 known ant species are spectacularly successful, being found on every continent except Antarctica. Take the Argentine ant, a species with the largest recorded societies of any multicellular organisms. The imaginatively named 'large supercolony' (which may number over a trillion individuals in California alone!), covers 1,000km of the western United States, from San Francisco to the Mexican border, as well as 6,000km in Europe, 2,800km in Australia, 900km on New Zealand's North Island and growing areas on Hawaii and Japan. Remarkably, even though it stretches over multiple continents, it is a single society. The evidence is in the chemical make-up of the hydrocarbons on their cuticle and the way in which ants from different sites behave towards other ants. Take an ant from the colony in California and drop it into the heart of the same colony in Japan, and the Japanese ants will rub antennae with it and treat it as if it is one of their own. But take an ant from a different colony in California and drop it into the large supercolony (in California, Japan, or anywhere else) and the unfortunate creature will be ripped apart in minutes. The borders of colonies may be just a few centimetres apart, but if they are not part of the same society, they will be littered with dead bodies where the two sides have tried to encroach on the other – they are fiercely territorial and fight voraciously to the death.

You'd be forgiven for thinking that ants are much of a muchness, and that although there are numerous species, they all pretty much look the same and do the same things. In fact, these little creatures exhibit a staggering diversity of form and function. Take the Madagascar Dracula ant – worth a mention for its name alone. Madagascar is the world's fourth largest island, and a biodiversity hotspot due to its ancient origins as a fragment of the supercontinent Gondwanaland and the resulting evolutionary isolation since splitting from India some 88 million years ago. The Madagascar Dracula ant's particular peculiarity is their habit of feeding on the blood of their own larvae. Well, not 'blood' as we know it, but the blue-green insect equivalent called hemolymph. These ants scratch and chew holes into their unborn babies, sucking out some of the hemolymph for themselves or their queen. They don't do enough to kill the larvae, so the behaviour is called 'non-destructive' cannibalism. But still, it's pretty extreme.

Weaver ants also exploit their young but in a slightly less gruesome way. Their larvae produce sticky silk, which they use to build cocoons and the adults use to 'glue' together leaves as they build nests in the trees. Holding the larvae in their jaws, worker ants squeeze them like little glue bottles to exude drops of silk on to the leaves until the edges are all stuck together.

We could go on, but our focus here is on planning as it relates to food storage. While Aesop's grasshopper spent the summer singing in the fields, the ants worked hard stockpiling rations to last them through the winter – and, remarkably, plenty of evidence suggests that this is grounded in fact. Some ants are farmers, if we are happy to accept that the only reason they fall short of the dictionary definition is that they aren't people. And, perhaps even more remarkably, different species farm in different ways.

The Argentine ants, for instance, have a particular liking for 'honeydew': the sticky, sweet waste product of sap-feeding aphids. This is what collects on the roof of your car when you park under a leafy sycamore on a hot summer's day. Argentine ants 'milk' the honeydew from engorged aphids by stroking their antennae over the aphids' abdomens. This behaviour was observed by numerous naturalists in the nineteenth century, including Charles Darwin who tried, and failed, to mimic it:

> I removed all the ants from a group of about a dozen aphides on a dock plant, and prevented their attendance during several hours. After this interval, I felt sure that the aphides would want to excrete. I watched them for some time through a lens, but not one excreted; I then tickled them with a hair in the same manner, as well as I could, as the ants do with their antennæ; but not one excreted. Afterwards I allowed an ant to visit them, and it immediately seemed, by its eager way of running about, to be well aware what a rich flock it had discovered; it then began to play with its antennæ on the abdomen, first of one aphis and then of another; and each, as soon as it felt the antennæ, immediately lifted up its abdomen and excreted a limpid drop of sweet juice, which was eagerly devoured by the ant. Even quite young aphides behaved in this manner, showing that the action was instinctive, and not the result of experience.

These ants are, as we've already seen, feisty little creatures, and they voraciously protect aphids from ladybirds and other predators, in essence farming them for honeydew in an equivalent way to a farmer that keeps cows for milk. The consequence is that across vineyards and plantations in the US, the ants' presence has caused aphid numbers to boom, creating a severe nuisance for farmers.

Moving to the dry, desert regions of western America, Australia, Mexico, South Africa and Guinea, honeypot ants have taken things one step further. Their solution to the problem of surviving in extreme arid conditions is particularly ingenious, and like the Madagascar Dracula ant and weaver ant, also involves exploitation of their own colony. It starts during the rainy season when food is abundant and worker ants can gather large quantities of sugary nectar solution; enough, in fact, to easily last through the arid months. But where to store it? If we were storing a liquid we would jar or bottle it and leave it in a cupboard until we wanted to drink it. In essence, the ants do that too, but in the absence of bottles they have taken the unusual step of using other ants as their storage containers. Here's how it works. Honeypot ants, like many ant species, show a physical division of labour such that different ants are specialised to perform different roles within the colony. As well as having workers for foraging or colony maintenance, and soldiers for protection, honeypot ants have a unique class called repletes, and they're the storage bottles. Workers feed so much nectar to the repletes that their abdomens swell as large as a grape and they are literally immobilised. They become part of a vast living larder; a storeroom full of fat-bellied, ant-shaped honey drinks, heavily guarded against intruders, including humans, who also value the sugar solution. When food dries up outside, these giants are utilised for food. All a hungry worker needs to do is stroke the antenna of one of the repletes to stimulate regurgitation.

Impressive. But the real kings of cultivation actually grow things, and their behaviour is nothing short of remarkable. These are the leafcutters, a generic name given to numerous species of tropical ant living in South, Central and North America. Leafcutter colonies are also vast (a single colony can number over 5 million individuals and tends to take the form of several huge underground nests, connected by tunnels). Large soldier ants protect the colony, while at the other extreme tiny 'minima' ants tend to the colony's food supplies. 'Media' and 'maxima' ants are foragers and have powerful jaws, which they use to harvest pieces of leaf from the rainforest; they return to the colony in an orderly march along branches and down trunks, carrying leaf fragments that can weigh up to 20 times their body weight. The leaves aren't food; instead, the ants chew them into even smaller pieces and transport them deep within the nest to the 'fungus garden', where the decaying mulch encourages the growth of a type of fungus. This is the ants' food: the entire colony relies upon the output of these fungus gardens, which are diligently tended and guarded from intruders. So specialised is the system that the ants have even evolved a mechanism of protecting their fungus garden from unwanted microbes and parasites, by cultivating an antibiotic-producing bacteria on the surface of their own bodies! Researchers are now studying the ants to learn more about the bacteria and the antibiotic, which is related to an antifungal used in modern medicine.

Aesop's ants were almost certainly 'harvester' ants, a broad term used to describe several species that feed on seeds. Harvester workers forage for seeds up to 50km away from their nest, bringing them back to the colony where they are stored in underground granaries for later consumption. As outlined in the 'bible of myrmecology', *The Ants* by Bert Hölldobler and EO Wilson, the geographic distribution of these species around the arid zones of the Mediterranean and the Middle East make it likely that ancient writers encountered the harvester ants, particularly those of the *Messor* genus. Messor was the Roman god of agriculture, presiding over the crop harvests, so the genus name is apt. In their words: 'These middle-sized, conspicuous ants are often serious grain pests, and it is to them that the writings of Solomon, Hesiod, Aesop, Plutarch, Horace, Virgil, Ovid and Pliny almost certainly allude.' Bingo! Here, then, we have found Aesop's ants. Behaving as Aesop depicted, harvester ants work to store up grains that will help them survive during leaner times. This means that Aesop's ants are not just grounded in fact; they *are* fact. We could leave the fable there, concluding that Aesop got it right. But it's only part of the story; the really interesting bit is not whether the ants do it, but how.

★ ★ ★

In the late nineteenth century, individual ants were presumed to have rich emotional lives and intelligence, including the ability to plan for the future. As anecdotes about behaviour became less accepted, it became clear that basing conclusions about ant intelligence on the abilities of individual ants was not appropriate because they aren't akin to individual people, dogs or fish. Instead, ants demonstrate collective intelligence, each individual being one component of the single entity that is the colony. Each ant is a tiny contributor to the overall health and functioning of the colony, much like the individual cells in our body.

At the colony level, ants can achieve incredibly complex things. They can link together to form living, floating 'rafts' for crossing water, or great 'tug-of-war' chains that allow them to drag back prey that would be impossibly large for one individual. Through the efficient use of simple rules, the actions of individuals can sum together into the sophisticated collective intelligence of the colony, often termed a 'superorganism'. Take, for example, finding food. Workers leave the nest to forage for food in a random manner to begin with, wandering all over the place until they find something. Watch them for long enough though, and you will see them start to converge on the same path until it becomes a bustling highway of ants, scurrying business-like from food to nest in remarkable order. Such efficiency is not the result of coordinated planning, but merely the outcome of each ant's pre-programmed impulse to first, deposit pheromones when they have found food and are returning to the nest; second, deposit more pheromone when the food is of high quality; and third, to follow pheromone trails that they detect in their environments. Simple rules, but when followed by thousands of individual ants they result in a thickly scented ant 'highway', a self-reinforcing effect that results in efficient, rapid food discovery and transport.

Ant highways can teach us a thing or two as well, since ants are one of the rare species that collectively engage in two-way travel. We have likely all experienced the frustration of being sat in a traffic jam, or of needing to suddenly brake on a motorway for no apparent reason. Ants, in contrast, regulate their traffic flow under the most crowded conditions. In one study, researchers recorded the movement patterns of Argentine ants, whose colonies were connected to a food source by a bridge. To evaluate how ant density affected their movement, the team tested different sized colonies (ranging from 400 to over 25,000 individuals) and bridge widths. They found that, unlike in humans, traffic kept flowing even when the largest colony was trying to cross the narrowest bridge. In humans, flow starts to decline at road occupancy levels of 40 per cent, whereas for the ants flow was maintained at 80 per cent occupancy. They do it by proactively reducing their speed as the route becomes busier, meaning that they reduce the risk of time-consuming collisions. We could learn a thing or two from these ants, but we won't because we behave to maximise our individual goals whereas ants are acting to ensure survival of the colony. This isn't intended as a slight on the human race; as we heard previously, the unique genetic relationship in many eusocial insects means that they exhibit much higher levels of cooperation than other animals, humans included.

One of the reasons why ant societies work so well is that they divide up tasks: some individuals take care of colony defence, others go on foraging trips, others tend to the colony's larvae and so on. It's an efficient way to organise a group and some ants have taken it to extremes. The replete class of honeypot ants, for example, are physically specialised for that job, as are the big soldier leafcutters. Taking these ants out of their usual routine results in them behaving with very little apparent purpose. And because they aren't able to generalise their behaviour, they have lower survival too. One study found that isolated foraging carpenter ants lived for about six days, compared with 66 days in the colony. Finding food wasn't the problem (although it took longer and burned up more energy), but they hit a serious stumbling block when they tried to eat it. Under normal colony conditions, foraging ants would deposit the food at the nest for storage and communal use, but for these solitary foragers the food simply stayed in their crop – without the social stimulation from other ants they were unable to process it further.

In other species the division of labour is a bit more flexible, and this flexibility is thought to be one of the contributing factors to the spectacular success of Argentine ants. It depends on the number of other ants that are doing a particular job. The more ants that they encounter who are doing something, such as foraging or colony defence, the more likely they are to switch to doing that too. This enables resource to be diverted to where it's needed, which might give the Argentine ants an advantage. It might also explain why many colonies have a large group of 'lazy ants' doing nothing. Rather than being freeloaders, they are possibly just waiting to be needed.

Professor Nigel Franks established the Ant Lab at the University of Bristol in 2001. Together with numerous students and colleagues, his lab has pioneered the use of the rock ant as a model system in which to investigate ant behaviour. In nature, these tiny ants live in flat rock crevices; they prefer dark, enclosed spaces and often construct little walls to enclose the nest when they're not naturally formed. In the lab, rock ants can be housed in nests made from two microscope slides held apart by a cardboard perimeter; a nest can be made uninhabitable by simply lifting off the top slide. Using this simple system, Franks and colleagues have found that division of labour could happen as a result of individual changes, such as age, corpulence[*] or experience, and that individual ants may also have different thresholds for taking up a particular task.

Tandem running is a good example. This is something that several ant species do, including the rock ant, whereby one individual leads a single other ant to a new food resource or nest site. The 'follower' keeps track of the 'leader' ant's movements by repeatedly tapping her antennae on the leader's abdomen and legs, pausing every now and then to take stock of her surroundings. It's thought that these pauses enable her to learn landmarks that will help her to repeat the route when she goes solo. Franks was the first to study tandem running in detail and he showed that it fulfils the criteria of teaching, another behaviour once regarded as uniquely human. Follower ants find food faster when being led and the leader experiences a cost for the behaviour, as leading a naïve ant is about four times slower than proceeding directly to the food. When the follower stops tapping the leader also stops, thereby

---

[*] 'Chubbiness' or 'podginess' isn't the right word for an ant.

adjusting her behaviour to the naïve individual. Tandem running helps to bring several ants up to speed on the location of a new nest site so that they can transport eggs, larvae and the queen quickly and efficiently. It's not always necessary though. Once enough of the colony are in the new nest, it's about three times faster for teachers to switch to simply picking up and carrying any stragglers.

Ants are thorough when it comes to evaluating new nest sites, and evidence shows that they will go on reconnaissance missions to survey potential 'housing stock' in their neighbourhood before they need to move. One experiment in Franks' lab tested this by providing ants (that were safely housed with no need to emigrate) with the opportunity to explore a potential new nest site for a week. It's like you getting the keys to a vacant house a couple of streets away, even though you have a perfectly good house and aren't looking to move. Do you go and have a look anyway? After the week, the ants' current nest was made uninhabitable and they had the choice of moving to the site they'd had the opportunity to explore or to another, novel nest site that had identical specs. The team found that when the available options were as good as their original nest site, they relocated to them randomly. But when the available options were worse than their original nest site, the ants overwhelmingly chose the novel nest. In the same way, if the house that you checked out had a damp problem and a rat infestation, when your house unexpectedly flooded and you could only move to that or a 'mystery' house, which would you pick? The idea is that the ants investigated the other option and learned that it was less good than where they were currently living. Even though there was no immediate need to learn and remember this information, they did so, likely through a combination of landmark learning and pheromone marks that signalled 'this nest isn't worth exploring again'. When their current nest was destroyed, they were able to use this information to quickly decide, choosing the novel option even though it was identical to the one they rejected. It's called 'latent learning', meaning learning that takes place without an immediate reward, which is subsequently produced when reinforcement is available. Latent learning is what's going on when your dog, who you've been trying to train to sit nicely without any treats and who seems to be steadfastly refusing, does it perfectly the first time you offer a treat for the behaviour. The information was going in all along, it just needed some reinforcement to produce it. In the case of the ants, it seems to suggest planning, because their motivational state in the week before their nest was destroyed would have been different to what it was during emigration. In actuality, the ants were simply storing up the learned information until it was needed. In their write-up, Franks and team concluded that although their findings 'suggest that ants may also in effect anticipate the future ... planning for the future can be a social activity based on relatively simple rules without any form of mental time travel'.

Fact or fiction?

Aesop's fable of the ant and the grasshopper doesn't quite hold true. The grasshopper may play the role, although we reach this conclusion partly through the absence of evidence about grasshopper behaviour and partly because it was going to die anyway, come the winter. Harvester ants, on the other hand, do store up grain to last through leaner times, so Aesop's ants are partly based in fact. Where the difference comes is the mechanism governing this food storage: an evolved collective intelligence, rather than each ant thinking ahead. There is evidence for latent learning in ants, but this is a simpler form of planning that does not require mental time travel. The ants would not have derided the grasshopper for his lack of forethought, as there is no evidence that they possess such a concept themselves.

What can we switch in to take the place of the ant? Remember that Wolfgang Köhler set out the conditions under which to identify future planning almost 100 years ago, concluding from his observations that chimpanzees were limited to the present. It took decades before those studies were done, but if Köhler was alive today it is possible he may change his mind.

Other notable researchers have also changed their mind over the past 20 years. Psychologist Bill Roberts, professor emeritus at the University of Western Ontario, Canada, published an article in 2002 entitled 'Are animals stuck in time?' In it, he answered the question with a definitive 'yes'; there was no evidence at that point to refute the idea. Ten years later, however, Roberts had revisited his position – in a 2012 publication entitled 'Evidence for future cognition in animals', he emphasised the number of field and laboratory studies that had been conducted on the topic in the intervening period. He described evidence from experimental studies of apes, rats, black-capped chickadees, monkeys and scrub jays that all suggested mental time travel, noting that although various 'lower-level mechanisms' had been suggested that could theoretically account for the behaviours, in his opinion the most parsimonious interpretation was that future planning was not restricted to humans.

Future planning relies on two key abilities. First, self-control. You need to suppress a current desire for a reward and that's hard. Second, you need to have a detached mental experience about a future or past event. Both abilities develop in us at around the age of three to five years old although, it must be said, we don't exhibit them all the time. How many of us have ever bought a slab of chocolate and opened it with the intention of eating one square and saving the rest? How did that work out for you? We can do it, sure, but that doesn't mean that we always will, because we don't always behave as perfectly rational creatures. Being able to override our current desires depends on various things – our emotional state at the time, the rarity of the commodity in question, what other people are doing and so on.

Similar factors come into play in other animals too. Capuchin monkeys aren't very good at self-control, as evidenced by a task where they could wait to exchange a small cookie for a larger one and were only able to wait for up to 20 seconds. Apes are much better at this. In a similar cookie-exchange task, most chimpanzees waited for an impressive eight minutes to trade in the small cookie, but only when the larger reward was 40 times bigger than the one they had in their possession! In another task where pieces of chocolate were counted into a bowl in front of them and delivery of the pieces stopped as soon as they began eating, chimpanzees and one orangutan waited patiently for three minutes for the maximum of 20 pieces to be placed before consuming them.

Observational reports suggest that chimpanzees are also capable of thinking ahead. There's the male who cracked open a tortoise and feasted on half of it, then stashed the rest up a tree and returned to finish it off the next morning. Then there's Santino, a wily male chimpanzee who is infamous for his antics at Furuvik Zoo. Early in 1997, after three years living relatively peacefully in his moat-bound enclosure, Santino started to display typical male dominance behaviour towards people who stopped to watch him across the moat, including hurling rocks at them. Subsequent sweeps of Santino's island compound revealed five hoards of stones, all taken from the moat and located at the shoreline nearest the public viewing area. A baffled member of the zoo staff set up a hide to watch and was astonished to see Santino collecting stones from the moat and piling them together in little heaps, creating an arsenal of weapons that he could hurl. What was even more remarkable was that he collected the stones not when the visitors arrived and he became agitated, but much, much earlier: in some cases, first thing in the morning before the zoo had even opened. Since this initial discovery zoo staff have removed hundreds more stashes of stones to protect their visitors.

Researcher Mathias Osvath was already working at the zoo, collecting data for his PhD on chimpanzee cognition, and he decided to conduct a side project into chimpanzee and orangutan planning abilities. 'I thought that I will just show that apes cannot do this, because that was the assumption,' he tells me. 'I showed them, here is the apparatus. But you have no tool, so what will you do tomorrow when you get the opportunity to get the tool? And they just passed. I thought, OK, maybe this is something basic, but they won't pass a self–control task – I was certain about that.' He presented the animals with a choice between a tool (that could be later used to obtain a highly prized food reward) and a smaller, but still valued, treat (a grape). They chose the tool. One objection that is often raised in these kinds of studies is that after the repeated use of a particular tool, it might end up having a higher reinforcement value than the food and so picking the tool may act to satisfy a current desire for a valued object, rather than any form of future planning. To control for this, Osvath ran another test where, once the animal had selected the tool and had it in its possession, they were immediately offered another choice between the same tool again or a piece of fruit. If they simply valued the tool then they'd be expected to pick it again, even though they already had one in their possession. If, on the other hand, they were anticipating the future usefulness of the tool then they would be expected to pick the fruit, because having another tool wouldn't be any more useful to retrieving the reward. The results were conclusive – all animals picked the fruit on every trial. In the final control, the chimps were presented with novel objects, one of which could be used to obtain the treat in the future, two that were 'distractor' objects, with no function to the task, and one familiar tool – a stick that the animals regularly used to dip for honey – that was useless to the future task. If the chimps were choosing a tool because they associated it with food, they should choose the dipping stick. Instead, they showed a strong preference for the novel, functional tool. Osvath had to rethink his position: 'I became convinced that this type of planning, making a decision for the future, is most likely not uniquely human.'

When Osvath started his experiments on planning at Furuvik Zoo, Santino was not his focus; he was the dominant male in the troop and refused to participate in any experiment, so Osvath worked with the troop's females instead. In any case, he was expecting to demonstrate that the animals lacked this ability, so he brushed aside the anecdotes about Santino he was hearing from zoo staff. 'The zookeepers were telling me, "Of course they can plan", and they told me about Santino. And I was very, very sceptical.' But, as Osvath's experiments started to generate positive data, he began to change his mind. He went back through the zoo's logbook and observed the ape for himself, including recording Santino's apparent mental state throughout the day. Osvath found that in the early morning, when Santino gathered and stored the stones, he appeared calm. After the zoo had opened to visitors, sometimes several hours later, Santino started to show signs of agitation. His behaviour advanced in complexity when, as well as collecting rocks from the moat, he started to break off small chunks of concrete from a structure in his enclosure and add these to his arsenal. By this point the zoo guides had become wise to Santino's moods, so when he started to whoop and jump they shepherded visitors to a safe distance. When the visitors retreated, Santino calmed down and stopped throwing stones. That wasn't the end of the story. Osvath went on to record Santino start to hide his stone piles underneath hay or behind objects. His most intriguing observation was to follow. Santino stopped aggressively displaying towards the visitors and, as a result, the guides stopped shepherding people away quite so quickly. The wily animal took the opportunity to move closer to the moat and to then hurl his hidden missiles without any warning. The zoo guides had learned, but so had Santino: a fascinating battle of wits between man and chimp.

Santino's decade of antisocial behaviour is extraordinary, but it's impossible to draw conclusions about chimpanzees or great apes from observations of one animal. What's more, since his behaviour occurred spontaneously, rather than during a controlled experiment, it is impossible to know how it developed. So, how do we test for future planning in animals?

Eminent experimental psychologist Endel Tulving proposed a thought experiment called the 'spoon test'. Imagine, Tulving said, that a girl went to a party and was told that she couldn't have any pudding because she didn't bring a spoon. If she was able to think ahead, then the next time she went to a party this memory of disappointment should prompt her to bring a spoon.

It often happens in science that different groups end up investigating the same question in parallel – and this was the case for mental time travel in animals. While Osvath was gathering data on the apes at Furuvik Zoo, Nicky Clayton was demonstrating evidence of flexible memory in jays (more shortly) and comparative psychologist Josep Call and his student, Nick Mulcahy, based at the primate research facility of the Max Planck Institute for Evolutionary Anthropology in Leipzig, Germany, had devised an experiment to test apes on an analogous version of Tulving's task. They asked whether orangutans and bonobos, our most distant and closest ape relatives, respectively, could save tools for future use, theorising that this may be more cognitively challenging than saving food, because it is one step removed from the goal.

The apes were presented with one of two different tasks in a chamber that they could see but not enter. In one task, there were grapes inside an open-ended tube, which needed to be pushed out with a stick; in another, a small flask of grape juice was suspended out of reach and needed to be dragged in with a hook. The chimps were then presented with tools in the absence of the tasks and after a one-hour delay they were allowed into the testing arena, where one of the tasks was set up. The animals stored tools on 70 per cent of trials and they were choosy, preferring to store tools that they had learned were more functional, even though the task wasn't visible at the time of choosing. Incredibly, some individuals even saved tools for the task when the interval was extended to 14 hours and included sleeping time. Importantly, on a control task where the apes could select tools, but when they went back to the testing arena no apparatus was set up, they stopped bringing tools in.

The future planning behaviour of great apes has been called 'domain general', because it takes place in different contexts – both tool-related and socially. Monkeys don't do this, and as a result future planning is considered by some to be a unique ability of the evolutionary branch that includes apes and humans. Yet apes aren't the only contenders vying to take the place of Aesop's ants. Sidling into view in that oh-so-typical way are the corvids, who are systematically demolishing every 'ape-centric' barrier that has been placed in their way. I had wondered whether my star bird, Pierre, was planning ahead by stashing tools, and while the corvid equivalent of this experiment has not yet been done, there is intriguing evidence for mental time travel in other members of the group.

The western scrub jay was introduced earlier for its place in the history of corvid cognition research and no discussion of mental time travel would be complete without it. Remember that western scrub jays, like many corvids, cache excess food in autumn. And while we already dismissed the idea that caching per se is indicative of any form of future planning, animals that store food, like the western scrub jays, make particularly good subjects for examining the cognitive components of food storing. Food that has been stored needs to be recovered, which may necessitate remembering the location of hundreds or thousands of hoards. And that means taking a step in the other direction and asking what animals know about the past as well as the future.

There are different ways in which things can be remembered. It was Tulving who, in 1972, proposed the existence of three memory types, and while his proposal has been refined it remains influential. Tulving noted that each memory type enables the individual to make use of acquired knowledge. But they differ, he wrote, in the type of knowledge processed and the way in which the different kinds of knowledge are used. Take riding a bike. The first type of memory is called procedural; it's why, once you've mastered the basics of riding a bike, you can jump on and pedal away without consciously thinking about when and how to move your legs or shift your weight for balance. Then there's your semantic memory, which deals with the storage of facts, but which do not necessarily have any personal connection to you – for example, you may remember the name of the last Tour de France winner, but may not necessarily remember when or how you learned that. Finally, there's your episodic memory, which is the personal, subjective memory that allows you to relive a specific event in your mind. Perhaps many years ago you went to watch a stage of the Tour and you can bring to mind your experience of standing at the road edge, feeling the sun on your face, and hearing whistles and cheers as the riders passed. Tulving proposed that the three memory types were hierarchical: that it would be impossible for an organism to possess episodic memory without possessing semantic, and that it would be impossible to possess semantic memory without possessing procedural. Tulving suggested that episodic memory went hand in hand with a uniquely human type of consciousness (called 'autonoetic'), which was essential to allow the re-experiencing of past events.

Episodic memory may be subjective and unique to each individual that experiences it, but at its core is an ability to remember the 'what, where and when' of a specific event. It's how that plays out in the mind that is inaccessible, so trying to ask whether animals, that cannot describe their experiences, possess any form of episodic memory is therefore an impossible task. But what we can do is assess whether they remember specific components of their past behaviour and this is where the scrub jays come in. In a groundbreaking experiment published in 1998 and entitled 'Episodic-like memory during cache recovery in scrub jays', Clayton and psychologist Tony Dickinson showed that the birds formed specific what–where–when memories of caching events, and they then made use of these memories in a flexible way. In the study, western scrub jays could cache two types of food in different locations: preferred, but rapidly perishable, wax moth larvae and less-preferred, but non-perishable, peanuts. The birds were provided with the opportunity to recover the food after a short (four-hour) or long (124-hour) interval. Clayton and Dickinson found that all birds preferentially recovered larvae after a short interval, but after the long interval the birds' prior experience affected their choice. Those that had previously experienced worms rotting switched to recovering peanuts, while those that hadn't kept choosing worms. Western scrub jays therefore successfully recalled the type of food they had cached (what), its location (where) and how long ago it had been stashed (when).

We cannot ask the jays if, when they are recovering previously stored food, they are re-experiencing the memory of a previous time when they left some grubs for too long and came back to find them rotten. And because of this, Clayton and Dickinson termed the scrub jays' memory as 'episodic-like'. Since then, evidence for episodic-like memory has been reported in other birds, mammals and even cuttlefish; yet the scrub jays remain the top example that an animal may possess this type of memory, and the jays' proficiency in these tasks prompted Clayton and team to ask if the birds could also travel forwards in mental time, with a series of cunning experiments centred around planning for 'breakfast'.

In the planning experiment, each of the eight birds tested had access to three adjoining chambers, which could be divided from each other or left open as one large area. For six days, each bird was randomly shut into one of the end chambers in the morning. One end was always provisioned with food, in the form of powdered pine nuts, while the other was always empty, meaning that sometimes the birds got breakfast and sometimes they didn't. For the rest of the day, the birds could move around all three chambers and powdered pine nuts were freely available. On the evening of the seventh day, the powdered food was replaced with whole pine nuts, and the birds were able to store food for the first time. The team watched eagerly. If the birds could think ahead, they should preferentially store food in the 'no-breakfast' chamber, for the possibility that they may be shut in there the next morning. If they weren't planning ahead, they were expected to stash food at random in both the 'breakfast' and 'no-breakfast' chambers. The birds hid more than three times more pine nuts in the 'no-breakfast' chamber compared with the 'breakfast' chamber, indicating that they were pre-empting the lack of breakfast the next morning. It's analogous to Tulving's girl who, remembering the disappointment of not being allowed pudding, brings a spoon to the next party.

Critics pointed out that the birds might show the same behaviour if they had simply learned to associate feelings of hunger with one chamber, without the need for any future thought. Another experiment took place, using an identical method, except this time the birds learned that in the morning one chamber always contained peanuts and the other always contained dog kibble; this way, the birds were never hungry. On the evening of the seventh day, when provided with the opportunity to store either food item, they chose to stash more peanuts in the 'kibble' chamber and vice versa. Jays, like all corvids, are generalist omnivores, so this suggests they may have been thinking ahead to ensure they had a variety of foods available in the morning.

These are pretty convincing results. But critics of the jay studies cite the very point that Clayton had initially exploited – because storing food is a typical behaviour for the jays, their impressive abilities may reflect specific adaptations for food caching, rather than the type of domain general planning indicated for the apes.

Back to Osvath, who continues to work with chimps, but who was immediately inspired by Clayton's initial findings to work with corvids too. He initially tried rooks, but found them to be 'completely unworkable', so he turned instead to ravens. To avoid the criticism levelled at Clayton's studies, Osvath and student, Can Kabadayi, devised some experiments with ravens that involved atypical behaviours. First, a task in which the birds needed to learn to barter with humans; second, a task in which they needed to learn to use a tool. Neither is a natural behaviour for ravens, though they can readily learn both. And once they'd learned that a stone could be used as a tool on an apparatus, and that they could exchange tokens with people for food rewards, the ravens' behaviour was at least as impressive as the apes. In both tasks, after the ravens had experienced one trial where they were unable to obtain food because they had no tool or token (i.e. akin to Tulving's girl), they selected the functional object from a given array significantly more often than predicted by chance, including from the first trial. They then successfully used the object after delays of up to 17 hours. Kabadayi and Osvath also tested the birds' self-control, offering an immediate (but smaller and less valued) reward at the same time as the object array. They found that when there was a 15-minute delay before the tool or token could be used, the ravens went for the immediate reward about 25% of the time. In contrast, when there was no delay, all birds chose the tool or token on all trials, ignoring the immediate reward. The study suggests that ravens are domain-general planners, thinking not just outside their current motivational state but outside their evolved behavioural predispositions. As Kabadayi and Osvath concluded, the behaviour of these 'avian dinosaurs' was strikingly similar to that of apes, providing further evidence for their remarkable evolutionary convergence.

Osvath and Kabadayi elaborated on their findings in a paper entitled, 'Contrary to the Gospel, ravens do plan flexibly'. And contrary to Aesop's fable, ants do not. Yet deciding on a replacement for the ants is tricky – apes and corvids are both strong contenders, so is there any way to tell them apart? Osvath thinks so: 'I've seen apes doing it, in several studies. But when I see the corvids do it, it's such a difference. You can really see that they get it.' It's not a consensus, of course, and ongoing studies will provide more answers, but for now the corvids have got it. Our updated fable is 'The Raven and the Grasshopper'.

CHAPTER NINE

The Hare and the Tortoise

A hare once ridiculed the short feet and slow pace of the Tortoise, who replied, laughing: 'Though you be swift as the wind, I will beat you in a race.' The Hare, believing her assertion to be simply impossible, assented to the proposal; and they agreed that the Fox should choose the course and fix the goal. On the day appointed for the race the two started together. The Tortoise never for a moment stopped, but went on with a slow but steady pace straight to the end of the course. The Hare, lying down by the wayside, fell fast asleep. At last waking up, and moving as fast as he could, he saw the Tortoise had reached the goal, and was comfortably dozing after her fatigue.

In 2016, a video of a 'Real-life Aesop's fable!' surfaced and quickly went viral across social media. It showed a kind of re-enactment of 'The Hare and the Tortoise', although the animals were clearly in an arena and a rabbit was used instead of a hare. The race begins and the rabbit hops along for a bit, well ahead of the tortoise, but about halfway down the track it pauses and sits there, whiskers twitching, refusing to be enticed by its owner. The tortoise, on the other hand, plods away, straight down the track and past the rabbit, over the finish line where rapturous crowds burst into applause. There is clearly a reasonably sized audience for this event. Media outlets picked it up and triumphantly claimed that Aesop's fable was grounded in fact – tortoises do beat hares!

We are all familiar with this one, arguably the best known of Aesop's fables. The moral, 'slow and steady wins the race', provides a reassuring social message about the benefits of hard work and persistence over arrogance and shoddy workmanship. It is a message that has resonated with cultures over the ages and is today applied to all sorts of different situations. Plug 'tortoise and hare' into an academic search engine, for example, and you retrieve a long list of publications including, but in no way limited to:

> 'Tortoise or hare: Will resprouting oaks or reseeding pines dominate following severe wildfire?'
> 'Postal surveys versus electronic mail surveys: The tortoise and the hare revisited'
> 'The marathon of the hare and the tortoise: implementing the EU Water Framework Directive'
> 'The tortoise and the hare: is there a place in spine surgery for randomised trials?'

It seems that we're all on board with the moral of this fable: that contrasting strategies can have different outcomes, and that it may sometimes be better to take the slower, more considered approach than the apparently quicker alternative.

But from a zoological standpoint, does anyone really believe that a tortoise could ever beat a hare?

The answer is very much 'it depends', which is often the most truthful yet least satisfying of all answers. By which I mean it depends on the race.

★ ★ ★

Let's address the obvious upfront. If a tortoise and hare could somehow be encouraged to sprint at top speed down a short, straight track, the hare would win hands down. The European hare is the fastest land mammal in the UK, capable of reaching speeds of around 70km/h. That's almost 20m every second, meaning that in 'one-oh-oh' the hare is not only up and out of the blocks, it's a fifth of the way down the sprint track. Hares make Usain Bolt look slow. Tortoises, of course, don't come anywhere close to these speeds. The red-footed tortoise has been described as 'lively' because it can travel 85m/h, which scales up to about 2km in a day. In his observations of giant Galapagos tortoises, Charles Darwin recorded one large animal walking at the impressive pace of 55m in 10 minutes, which if maintained for a day (with a little time for eating on the way) would be a cool 6.5km. The tortoise world speed record is held by a young leopard tortoise called Bertie, who in 2015 absolutely obliterated the existing record, held since 1977 by Charlie, of 43.7 seconds. Bertie charged up the 5.5m custom-built track in a searing 19.59 seconds, translating into an impressive top speed of 0.28m/s. That's in the region of 70 times slower than the hare.

Aesop did well in pitting against each other two familiar yet staggeringly different animals: tortoises and hares are worlds apart. To put it another way, they're 320 million years apart; that's when the last common ancestor of mammals and reptiles is known to have lived. In order to examine what each of these animals is capable of, the first question therefore needs to be, 'how did their respective evolutionary histories shape the tortoise and the hare?'

We'll start with the similarities and for that we need to go back – way back. Tortoises and hares, like us, together with all other mammals, birds, amphibians, reptiles and fish, are vertebrates – that means we all have vertebrae that surround a central nervous system, which is controlled by a brain. Simplifying things hugely (and skipping over hundreds of millions of years of evolution in the earliest animals), the first vertebrates were early, jawless fishes. All other groups are descended from a lineage of jawed fish that, somewhere between 400–350 million years ago, evolved fleshy lobe-like fins that enabled them to push themselves out of the water and subsequently drag themselves on to land. They passed on to their descendants the characteristics of lungs, paired limbs and a backbone, traits that enabled them to start colonising the land. All terrestrial vertebrates belong to a group called the tetrapods (meaning 'four limbs') and amphibians were the earliest; a group whose name essentially translates into 'living a double life', because the young live in water (e.g. tadpoles), while the adults live on land. Because of their reliance on suitable water sources for breeding, amphibians have been unable to exploit all terrestrial niches; nonetheless, we should remember that for tens of millions of years they dominated vertebrate life on land, a sobering thought given their perilous conservation status today. Things started to change when, around 340 million years ago, a branch of the amphibia evolved into the first reptile-like animals, whose scaly skin and semi-permeable eggs better adapted them for life more fully on land. There were two major developments early on in their evolution. First, they fully severed their ancestral dependency on water for reproduction with the evolution of amniotic membranes in their eggs; the embryos of this group (the 'amniotes') were sealed in their own, mini aquatic environments. Soon after, they diverged into two major groups called synapsids and sauropsids. What's remarkable is how these two groups went on to evolve.

The sauropsids include all the reptiles we're familiar with today, as well as the dinosaurs and many others long extinct. The dinosaurs were particularly successful, dominating the planet for around 180 million years. But even they couldn't survive the K–T extinction event,* at least not in that form. Some 66 million years ago, an asteroid crashed into Earth resulting in the extinction of three-quarters of all animal and plant species, including the dinosaurs, pterosaurs and large marine reptiles. Three major groups of reptiles survived to the present: *Chelonia* (turtles, tortoises and terrapins), *Archosauria*, including the *Crocodilia* (crocodiles, alligators and their relatives) and *Aves* (birds); and *Lepidosauria*, including the *Rhynchocephalia* (two species of tuatara) and *Squamata* (lizards, snakes and the little-known group of legless *Amphisbaenia*).

There were winners and losers in this life-changing event. As the dust settled, flowering plants emerged as the dominant plant group, diversifying into the stunning array of species we see today. And one group of dinosaurs made it to modern times, or at least, their descendants did. Birds belonged to a group of bipedal, flesh-eating dinosaurs called the theropods, which included velociraptors and *Tyrannosaurus rex*. Most of the theropods, being larger than 5kg, went extinct, but birds were already well adapted to survive the meteor strike. Feathers, warm-bloodedness and flight had already evolved tens of millions of years earlier, alongside a protracted period of 'miniaturisation'. These little flying reptiles took over the empty skies and subsequently evolved into the spectacular range of birdlife that we're familiar with today. That's right – birds are not only descended from dinosaurs, they're

---

*K–T is an abbreviation of Cretaceous–Tertiary, the geological eras either side of the event. Modern terminology refers to the K–Pg, or Cretaceous–Paleogene.

descended from the most terrifying, carnivorous group. Remember that next time you salute Mr Magpie.

And the synapsids? Well, early on they took a different evolutionary path. There have been many stages in their evolutionary history and each has set them further apart from their reptile-like ancestors, but a key one is that, like birds, they became warm blooded. Around 252 million years ago, a lineage of the synapsids called therapsids evolved the ability to regulate their body temperature; these were the earliest ancestors of mammals. Mammals lived alongside the dinosaurs, but they were a minor group of small, insectivorous, nocturnal creatures, prevented from moving into other niches by the dominant reptiles. But they too survived the K–T extinction and once life began to reform, therapsids found a world of opportunities. They underwent an extraordinary evolutionary radiation to give rise to the mammals that we know today, as well as many others that are long extinct.

Mammals and reptiles share a few commonalities in basic form and function. Nonetheless, 320 million years of divergence is long enough for some substantial differences to have evolved. And underpinning them all are radically different physiologies. We're all familiar with the fact that tortoises, as reptiles, are cold-blooded while hares, as mammals, are warm-blooded. These are the terms we learned as children and they provide a neat categorisation of the animal kingdom: amphibians, reptiles, fish and invertebrates form one group, while birds and mammals form the other. Yet, saying that a lizard is cold-blooded is a little misleading, because it gives the impression that these animals are not only cold inside, but that they're also not dependent on heat. In fact, every animal on the planet needs a way of harnessing heat energy – it is essential to drive the chemical reactions that underpin life. All animals can be more accurately described according to two axes: whether or not they generate heat and whether or not they regulate their body temperature.

Once reframed like this, there are better descriptive terms. Animals that generate internal, metabolic heat are called endotherms (*endo* meaning 'inside'; *therm* meaning 'heat'), while animals that don't are called ectotherms. Animals that maintain their body temperature independent of the environment are called homeotherms (*homeo* meaning 'same'), while those that don't are called poikilotherms.*

Hares, like us and most other mammals, birds and some fish, are homeothermic endotherms. We burn fuel, which comes in the form of the food we eat, to produce heat. An 'internal furnace' analogy is often used, but that gives the erroneous impression that there's one part of the body that is the 'heat-generating' part. In fact, it's happening in all of our cells and more specifically, the mitochondria that are packed into each cell; when we talk about our metabolism we're really talking about the mitochondria devouring glucose, producing energy and heat as 'waste' products.

What's more, we regulate it. There's a good chance that your current body temperature will be somewhere between 36.5–37.5°C, and that's not just a human quirk. All mammals and birds maintain a core temperature within a narrow set of limits, because the consequences of body temperature rising or falling too far or for too long outside of these limits can be lethal. Too low (below 35°C) and we enter a state of hypothermia; if left unchecked that can progress to organ failure and death. Too high (above 38°C) and we develop a

---

*In most cases, animals that generate heat also regulate it, so most endotherms are homeothermic and most ectotherms are poikilothermic – but nature never allows for such neat classifications. Some endotherms can be poikilothermic when they hibernate, because their body temperatures can fall to within 1°C of the ambient temperature – bears, bats and gophers fall into this group. Conversely, some ectotherms can be homeothermic, which they achieve by being huge ('gigantothermy', such as the dinosaurs), living in extremely stable environments such as deep ocean abysses, or behaviourally.

fever; above 40°C and things get dangerous as many life-sustaining enzymes stop functioning properly – and it's no exaggeration to say that without enzymes we simply couldn't exist.[*] Essentially, we have a window of about 5°C within which we can safely operate, which, given that we can move between an icily air-conditioned office and a scorching summer's day without batting an eyelid, means our bodies are doing a fantastic job of maintaining the norm.

We do this through a process called 'homeostasis', which is so important that we have a dedicated control centre in the brain; when the hypothalamus detects changes it activates multiple processes to bring our temperature back to normal. If we're overheating, tiny blood vessels in our skin expand to allow more blood to reach the skin's surface (called 'vasodilation'); we also produce sweat, which evaporates at the skin's surface, helping to carry heat away from the body. Many animals cannot sweat so they pant instead, rapidly expelling hot breath and drawing in cooler air. Other animals have evolved different, sometimes bizarre ways to cool down: storks and other birds poo all over their legs; hippos and pigs wallow in mud; kangaroos lick their forearms and certain birds engage in 'gular fluttering', vibrating the muscles and bones in their throats.

As the temperature drops, the regulatory centre sends out different signals. Metabolic rate increases, meaning the mitochondria consume glucose and fats faster to increase heat production. The little blood vessels in our skin constrict

---

[*] The human body contains around 75,000 of these proteins, each one controlling a specific type of reaction and with an optimum temperature at which it works best. The many chemical reactions that go on in our cells, such as converting glucose to energy, breaking proteins into peptides or harnessing oxygen, would take years in the absence of enzymes. Enzymes speed up reactions by over a million times, allowing them to take place in a fraction of a second!

('vasoconstriction'), reducing the amount of blood that is close to the skin's surface, while tiny muscles in our skin contract, pulling arm and legs hairs erect and trapping an insulating layer of air. Our larger skeletal muscles are activated to make mini contractions (shivering), generating energy and heat. Behaviourally, we are prompted to put on more clothes and actively seek out sources of heat, such as radiators or fires.

Homeothermy means that endotherms can live in the Arctic or the tropics, and they don't need to warm up to be active: a hare sheltering on a freezing winter night can jump into action as quickly as it can on a hot summer's day. The big cost is energy consumption: maintaining body temperature within a narrow set of limits uses the vast majority of an endotherm's consumed calories, particularly in the smallest animals that have the highest body-surface-area-to-volume ratio. Consider the beautiful little bee hummingbird: at 5–6cm in length and weighing approximately 1.6g, it is the smallest bird and among the smallest of all endotherms. And it suffers a heavy burden for the title, with the highest metabolic rate relative to body mass of any animal on the planet. With a heart that can beat more than 615 times a minute, the bee hummingbird must ingest more than half of its own body mass every day and up to eight times its body mass in water, meaning it's capable of pollinating 1,500 flowers every day. It, together with the tiniest mammals, the bumblebee bat and Etruscan shrew, is at the very limit of what is physiologically possible for an endotherm.

Our friend the tortoise is a poikilothermic ectotherm, like all reptiles, amphibia and most fish, which means it doesn't generate or regulate body heat. As a result, it can tolerate a much wider range of body temperatures. The world's tiniest animals are all ectotherms and, in fact, because they're reliant on exchanging heat with their environment, it is their maximum rather than minimum body size that is limited: ectotherms cannot be too large or too round because it would take too long to heat up or cool down. That's why the largest ectotherms tend to have an elongated or flattened body shape; with more of their body surface in contact with the air there's more opportunity for heat exchange.

Ectotherms aren't bound by internal metabolic needs, meaning that after a big feed, reptiles can go for weeks or months without needing to eat again. Because they produce very little metabolic heat, they regulate their body temperature behaviourally. First thing in the morning, for example, land-based ectotherms are slow and sluggish because their body temperature has cooled with the night air. To warm up, they move into the sun or on to warm rocks. This is why, in the UK, the best chance of spotting one of our six native terrestrial reptile species is to scour sheltered, sunny rocks, logs or even roads on a warm spring or summer morning.

All tortoises live in warm, semi-arid areas (there are no polar tortoises) and their activity patterns are strongly constrained by environmental conditions. In central Asia, for example, steppe tortoises are 'active' for only three months of the year, during which they move around for less than two hours per day. For the other nine months, these tortoises bury themselves in the sandy soil and enter a continuous period of inactivity to avoid the extreme heat and cold of their environment. Even more remarkably, they meet the total energetic demands needed for maintenance, reproduction and growth, as well as for the building of fat stores for hibernation, from foraging activity of just 15 minutes per day. This is in stark contrast to the pre-hibernation activity of an endotherm, such as a bear, which needs to gorge itself on as much energy-rich food as it can to build up sufficient fat stores. Hibernating bears can slow their metabolic rate to about 50 per cent, but will still burn through their energy reserves at a much higher rate than a hibernating tortoise, because their resting metabolic rate is so much higher.

The eighteenth-century naturalist Gilbert White inherited a tortoise called Timothy[*] from his aunt, and we learn much about his behaviour from snippets in his *The Natural History and Antiquities of Selborne*. In a letter to his friend and frequent correspondent, Daines Barrington, White summed up Timothy's methods for temperature control:

> Though he loves warm weather, he avoids the hot sun; because this thick shell, when once heated, would, as the poet says of solid armor, 'scald with safety'. He, therefore, spends the more sultry hours under the umbrella of a large cabbage leaf, or amid the waving forests of an asparagus bed. But as he avoids heat in the summer, so in the decline of the year he improves the faint autumnal beams, by getting within the reflection of a fruit wall; and though he never has read that planes inclining to the horizon receive a greater share of warmth, he inclines his shell by tilting it against the wall, to collect and admit every feeble ray.

While ectotherms are sluggish until they've heated up, relying on external heat doesn't necessarily limit an animal to life in the slow lane. Bearded dragons, the fastest reptiles, can reach speeds of up to 40km/h, leatherback turtles around 35km/h and black mambas over 20km/h. They're not up there with the hare, but they're not hanging around. Tortoises, admittedly, are not built for speed. They are, on the other hand, highly economical movers. One study encouraged tortoises to walk on a treadmill and measured their oxygen consumption; the researchers found that they were 'unusually efficient' compared with other animals. This, combined with their extremely slow speed, results in low metabolic costs associated with movement.

---

[*]Timothy turned out to be a female, but the name stuck.

Think of the tortoise a bit like a commercial cargo plane, while the hare is a fighter jet. Over a year, the former cruises around every day at a speed that maximises fuel efficiency. The latter displays bursts of high speed when it needs to, but is inactive for much of the year. This analogy was played out for real by a team of mechanical engineers, who in 2018 published a paper entitled 'The fastest animals and vehicles are neither the biggest nor the fastest over lifetime', in which they presented a theoretical model of maximum lifetime speed in animals and aircraft – entities that are not usually considered together. The team was particularly interested in 'outliers' that didn't fit the general pattern of bigger bodies being faster than smaller ones. Cheetahs, for example, reach higher top speeds than elephants, and tuna swim faster than whales. Using theoretical physics and plugging in a lot of data, they concluded that: 'When judged for speed averaged over lifetime, the fastest "sprinters" are in fact the slowest movers (as in Aesop's fable "The Tortoise and the Hare").'

The final point to make about speed is that two tortoises once moved at speeds of over 40,000km/h,[*] blowing the hare right out of the water. The pair of steppe tortoises, known only as 'Number 22' and 'Number 37', were aboard the Soviet Union's *Zond 5* spacecraft,[†] which in 1968 was the first to orbit the moon and return safely to Earth. They were strapped in for the six-and-a-half-day flight and remarkably, when the intact capsule was retrieved from the Indian Ocean, its occupants were not only alive, but in reasonable health! They'd lost about 10 per cent of their body weight and were hungry, but otherwise OK. At the current time, there are no

---

[*] According to NASA's website, a spacecraft leaving the Earth's surface needs to have an escape velocity of over 40,000 km/h.
[†] As well as some fruit flies, algae, bacteria and what has been reported as a creepy mannequin in the pilot's seat.

reports of hares ever going into space, excluding the green-furred cartoon character Bucky O'Hare.

★ ★ ★

Physiologically, the hare already has a huge advantage over the tortoise. For a fair comparison, the tortoise should at least be allowed to warm up and reach its optimal temperature, but even so – put them both on a track, force them to sprint and the hare's got it. Yet this is also the first point of contention. The sprint scenario presents what each animal is theoretically capable of, which is fine for studies of biomechanics and physiology, but not necessarily the same as what each animal might actually do. That's a different, and far more convoluted, story.

Let's start with the hare and the fact that hares aren't just 'big rabbits'. They're related, of course: hares, rabbits and their closest relatives, pikas (which look a bit like large hamsters, with round ears and no obvious tail), make up the evolutionary group called the lagomorphs. Lagomorphs are all important prey species, with annual mortality rates of 80–90% for some species. But they show a 'live fast, die young' strategy, meaning they have short pregnancies, fast-developing offspring and a short period of parental care; consequently, they can produce numerous litters of several offspring every year, balancing the losses from predation.

Hares and rabbits differ in some crucial ways. For starters, hares are much larger, with longer ears and hind legs, giving them more of a leggy lope than the rabbit's hop. Most obviously, rabbits live underground in communal burrows called warrens, while the largely solitary hare scrapes out shallow depressions in the ground called 'forms' to provide shelter while they rest during the day. Because of this, it's important that when baby hares (leverets) are born, they are fully formed and able to flee from predators. To achieve this, hares have longer pregnancies than rabbits and leverets instinctively scatter from their siblings, lying low in the same area and only congregating when their mother calls for them each night to feed. Rabbit offspring (kits), on the other hand, are born blind, hairless and in no way ready to function independently of their mother. Which is fine since they're cosseted away in a cosy underground nest. Here's the first possible context by which the tortoise could win. Anthony Caravaggi, lecturer in conservation biology at the University of South Wales, explains: 'Leverets, until they're fully weaned, will just hunker down in one place and wait to be fed. So, if it's a tortoise racing a very young hare, then it's going to win easily.'

Before the early twentieth century, lagomorphs were wrongly included in the same evolutionary group as rodents. It's easy to see why: both groups comprise furry prey species with large, continuously growing front teeth and similar behaviours. In fact, lagomorphs and rodents are thought to have gone their separate ways some 55–45 million years ago, so while they are sister groups, they've been long estranged. For context, the last common ancestor of chimpanzees and humans is thought to be just 5–7 million years ago, which was plenty of time to evolve some substantial differences.

Some 273 different species of lagomorph are represented in the fossil record, meaning that this group was once considerably more diverse than the present. Today, there are 91 species and they're split pretty much equally between rabbits, hares and pika. The hares are the most diverse and include the North American 'jackrabbits', which are not rabbits at all. They exhibit a lot of variation between populations, which has resulted in the field of hare taxonomy being fraught with controversy. The IUCN, for example, recognises 15 subspecies of mountain hare (the Irish hare is one, as are the mountain hares of the Arctic and Greenland, Russia, Japan and northern Europe) and 15 subspecies of European hare (across Russia, Europe and some of the Middle East; they've also been introduced in New York and Connecticut). Geographical separation is the first step on the road to speciation, but for now all are still classed as the same species.

The mountain hare is the UK's only true native lagomorph, unlike the European hare and rabbit, both of which were introduced in the past 2,000 years. European hares are thought to have evolved on the plains of Eurasia or Africa, making them well adapted to open grassland. Surprisingly large animals, they measure up to 70cm in length, with thick golden-brown fur, large hazel eyes and black-tipped ears. They were almost certainly introduced to the UK by the Romans and subsequently outcompeted the mountain hare, restricting it to the Scottish Highlands, Southern Uplands and certain Scottish islands, where they browse heather moorlands. The same has happened in parts of Sweden and Finland, with mountain hares pushed back to the snow line, and Caravaggi's research suggests that a similar thing might be happening to the Irish hare too.

Hares exhibit all the adaptations of a typical prey animal. Their large eyes sit either side of their head, giving them almost 360-degree vision and enabling them to see predators approaching from the front, sides and even behind. They are well camouflaged to their environments – European hares tend to live in arable lands, flattening themselves down in the earth during the day, while mountain hares undergo a remarkable colour change in the winter, turning from brown-grey to pure white to match the snowy Scottish landscape.* And those long

---

*Global heating is increasing the vulnerability of Scottish mountain hares – the hare's white winter coat increases visibility to predators on snow-free days, and a study published in 2020 found that hares are not adapting quickly enough to the reduced snow cover, with on average 35 more days each year where white hares are exposed on a snowless landscape. Yet, Caravaggi tells me that as long as the population size is healthy enough and predators don't key into this, the hares should be able to adapt – just as the Irish hares have done.

hind legs make them fast. Insanely fast. Both the European hare and antelope jackrabbit have been reported to reach speeds of 72km/h. But it's not all about flat-out sprint speed and, in fact, greyhounds are faster. Nonetheless, in the now illegal (in the UK) activity of hare coursing, two dogs were set against one hare and, incredibly, the hare often still got away. That's because what really matters is their manoeuvrability, and that comes down to biomechanics. Hares can accelerate at speeds of up to $4.4m/s^2$, which allows them to establish a lead over their pursuer. In many cases, that will be enough to deter the predator, but if it can keep up and gets too close the hare has another trick. With a top deceleration of $5.2m/s^2$, the hare can slam the brakes on even harder than it accelerated. If the predator can't anticipate that they will be left for dust as the hare jinks and rockets off in another direction.

Why did hares evolve to be such speed machines compared with their relatives? All are sought-after prey and all are likely to end up in the jaws of a fox or stoat unless they have a trick or two up their sleeve. For rabbits and pika, the trick is to dive underground at the first sign of danger. Hares evolved a different tactic, that of breakneck speed and agility, which ties in with their evolutionary history on the open steppes – in these environments, where there is little cover to hide, surviving means being able to shake off whatever is chasing you, which in turn selects for increased speed in the predator. In this evolutionary arms race, adaptations for speed in the prey drive the evolution of counter-adaptations in the predator in a cycle until an upper limit is reached whereby, for example, it's biomechanically impossible for either to get faster. It's why many of the world's fastest land animals evolved in open grassland environments – cheetah, pronghorn, springbok, lion. That hares are well adapted for this way of life is apparent from their anatomy and physiology. ECGs of their hearts, for example, show them to have thick ventricular walls that can pump out a high volume of blood on each beat (or 'ejection fraction'), resembling 'athlete's heart' syndrome in humans.

The energetic costs of flat-out speed can be substantial, so hares only sprint when they must. Otherwise, they rely more on evasion, moving through the landscape on familiar trails, ever vigilant to the threat of predators and hunkering down to avoid detection. Sometimes they employ a more bizarre tactic, purposefully making themselves *more* visible to approaching predators. This was seen in a study of European hares on the Somerset Levels, in which on 31 of the 32 occasions where a fox approached over open ground, the hare reacted by standing up on its hind legs and facing the fox, flashing its pale ventral fur. It seems maladaptive for potential prey to make itself more visible to a predator, but the idea is that by signalling to the fox that it's been seen, the hare is actually reducing the likelihood of it giving chase. Hares are about 50 per cent faster than foxes, so it's futile for the fox to attempt a chase once it's lost the element of surprise. The data agreed – no fox chased a signalling hare. It was a different story when the predator was a domestic dog – all hares either started moving away or adopted a crouching 'ready' position, primed for running. None of them stood up to signal their presence to these potentially faster pursuers.

When they're not being chased, hare activity levels peak at sunset and sunrise, and activity is lowest at about midday. Otherwise, they typically move and forage within their home range – the size of which correlates with how big the fields are and what time of year it is. There isn't a lot of information on hares' movement ecology, but we do know that they tend to move linearly around the landscape, getting between where they are and where they want to be in the most efficient way. Caravaggi explains that hares tend to follow familiar trails within their home range – not to the same degree as hillside sheep, but enough that when he goes out surveying he can anticipate which way they have crossed the landscape. What's more, they tend to be reluctant to leave their range, something that may work in the tortoise's favour. 'There might be [another] context in which a tortoise can beat a hare,' Caravaggi offers. I am all ears. 'Hypothetically, and this relies upon the hare being very stupid, which we don't know as I don't think anyone's tested their cognition either, if you had a starting point inside its home range and a finish point outside it, conceivably the tortoise might win because the hare might not want to leave its home range.' What's more, he adds, the route of the race may play a role: 'If the hare is unfamiliar with the route or if it doesn't conform to their preferred way of using the landscape, that might also go against it.'

Hare activity levels are seasonally dependent, peaking during the breeding season, which is typically between February and September. In springtime, testosterone levels surge in males, leading to intense competition called rutting. This is the best time to spot these elusive creatures, when their usual solitary tendencies dissipate and give rise to some seemingly erratic behaviour. Females dictate how breeding will play out. Coming out of winter when priorities were finding food or rearing the last litter, most females are ready to mate at the same time in early spring. To select the best male to father her litter, the female hare tests amorous males' ability to chase her over several kilometres. If the male cannot keep up, he is deemed unsuitable parent material and his advances are rejected. Even if he can keep up to the end of the chase, it's no guarantee of mating. If the female is not interested, she makes this clear by boxing his ears; a few well-timed punches putting paid to unwanted advances. Boxing hares are, therefore, usually not males competing with each other, but females rejecting males. As the hares stand on their hind legs while fighting, boxing is more noticeable in early spring than at other times of the year, but Caravaggi is keen to point out that hares can box all year round.

To casual observers, it's little wonder that the springtime behaviour of hares seemed borne out of madness and this likely inspired many of the popular representations of hares that we're familiar with today. The English saying, 'mad as a March hare' has been in general usage since the fourteenth century, with 'hare-brained' first recorded in the sixteenth century. And no character represents these better than Lewis Carroll's erratic, nonsensical March Hare. In mythology and folklore, hares are varyingly depicted as shapeshifters and tricksters. Some common themes are goddesses and witchcraft, transformations, fertility and the moon; yet it's hard to pin down a unifying theme, which seems right, given how hard it is to pin down a hare. They are shy and fleet of foot; you may never know a hare is crouched in the grass, watching you, or you may just catch the briefest shimmer of brown as it flees.

★ ★ ★

Fleeing from your pursuer is the most obvious way that prey animals maximise their chances of survival. But it's not the only option, as demonstrated by our friend the tortoise. He has his enemies, depending on where and what species he is; for example, in Loango National Park in Gabon, Central Africa, chimpanzees crack open the shells of forest hinge-backed tortoises by smashing them against tree trunks, while golden eagles have been reported dropping tortoises on to rocky ground in numerous areas of their broad geographic range. The ancient Greek playwright Aeschylus is said to have been killed by a dropped tortoise, with some claiming that the bird mistook his bald head for a boulder. Some of the other known predators of tortoises include bearded vultures, ravens, honey badgers, raccoons, coyotes and black bears.

As we've seen, tortoises didn't evolve for sprinting. They do dig tunnels,[*] which they use for nesting, hibernation and temperature regulation. But burrows are only useful refuges if you're quick enough to get to them. If not, you need another tactic, and the tortoise's is to tuck its head and feet inside its shell (aka 'carapace') and sit it out, waiting for the predator to lose patience and move on. It can flip the animal over, but that doesn't really help: the animal's underside is protected by another bony covering called the 'plastron'. The shell of a tortoise, like all living chelonians, is a marvel of bioengineering. Wholly and intimately connected to the tortoise, think of it less like a hard hat that sits atop the animal and more like a ribcage that somehow found its way outside the body and fused together. That's right, during their evolutionary history the turtle family (*Testudines*)[†] essentially turned themselves inside out and reassembled with their ribcage melded against their back and their belly. You can't just lift the shell off the body of a tortoise, in the same way that I can't just pull your arm out of its socket. Tortoise shells even inspired the ancient Roman military – 'testudo formation' was when shields were held above and front, so the unit was completely protected.

The other advantage of the shell, says Anna Wilkinson, professor of comparative cognition at Lincoln University, is that it allows the tortoise to carry on with whatever it's doing up to the point where it really senses danger. 'Otherwise, they really do just keep going.' That includes getting past any barriers that might be in their way, as Wilkinson recalls: 'I

---

[*] For this reason, it is important that pet tortoises are kept in enclosures with sunken perimeter fences.
[†] *Testudines* is the evolutionary order, which includes all turtles, tortoises and terrapins alive today, as well as extinct members of this group. *Chelonia* is the sub-order name given to the living (extant) members of this group.

once saw a giant tortoise over a series of months push down a barrier that was buried six foot into the ground. They persevere at things.'

Wilkinson heads up the cold-blooded cognition lab, where she studies how reptiles and amphibians perceive and learn about the world. As she tells me, for a long time 'everyone in cognition simply wrote reptiles off'. Some of that likely results from early experiments, such as those of the 'innovative' scientist Francesco Redi,* who in the seventeenth century is said to have performed several experiments on the longevity of tortoises. Redi took a direct approach, surgically removing the animals' brains and finding that they continued to move around and function for months. Going one step further, Redi cut the head completely off a large tortoise and found that, while the animal did not move around, it was sensitive to touch: if he pricked the area near the tail the headless neck and limbs were withdrawn into the shell. In another tortoise, 12 days after decapitation Redi cut open the animal and saw the heart still beating. While shocking and completely unacceptable by today's standards, in Redi's time animal welfare was not a concern unless the animal had religious or economic associations. Reptiles were considered to be primitive, lower forms of life, incapable of feeling pain and undeserving of human concern. As the naturalist Georges Baron Cuvier wrote in 1831: 'If, as every fact with which we are acquainted on the subject appears decisively to prove, the

---

*Redi was the first to disprove Aristotle's theory that 'lower' forms of life such as maggots emerged spontaneously from decaying matter – an idea that had been accepted for 2,000 years. Inspired by the approaches of Galileo Galilei and William Harvey, Redi studied this scientifically and found that maggots only appeared on meat on which flies had been able to land and on which they had laid eggs. Redi was also the first to come up with the idea of the controlled experiment, nothing short of revolutionary in the seventeenth century. His experiments with tortoises are described in William John Broderip's *Leaves from the note book of a naturalist* (1852).

intelligence of the animal is in proportion to the capacity of the cranium, the tortoises must be placed among the most irrational and inert of the living tribes.'

According to the progressive, linear view of evolution, reptiles were further down the ladder than birds and mammals. And, just as for birds, Victorian neuroanatomists found nothing to challenge this view. Reptiles also have a bit of a PR problem, which stems from their awkward, sluggish movements and their lack of facial expressiveness. We are far more likely to emotionally connect with animals that display familiar actions and expressions.

There were some attempts to study reptile cognition in the early twentieth century, including Robert Yerkes' study of a speckled turtle that could learn to solve a maze problem 'with surprising quickness'. But overwhelmingly the animals tested failed to solve the problems set, and by the 1950s and 60s, the resounding conclusion was that reptiles lacked intelligence. The characterisation stuck for decades. Wilkinson tells me how, during a talk on gaze following, the speaker discussed comparing mammals and birds, adding that obviously they couldn't compare reptiles. 'And I said, "Well of course we can compare reptiles," but it was absolutely dismissed.'

Wilkinson's interest in tortoises came about while she was working at a zoo in northern England. She started spending more of her lunch breaks in the park's reptile house, enthralled by the red-footed tortoises. 'What interested me was their sheer inquisitiveness and activity. They were always busy, moving around, interested in what was going on ... there was something about the red-footed tortoises that made me think, "That's astounding".' When she started her PhD, Wilkinson bought a young red-footed tortoise called Moses, and it's fair to say that he permanently shaped her career.

It was while attending a lecture about rats navigating through mazes that Wilkinson found herself thinking, 'Moses could do that.' Together with the lecturer, Geoffrey Hall, the pair set up a tortoise-sized maze equivalent to the eight-armed radial structure used for rats. At the end of each arm was a small piece of strawberry, a favoured treat. Moses could roam freely around the maze and had eight chances to get all the food. If he had no spatial memory, he would be expected to visit the arms randomly and take many attempts to obtain all eight strawberry pieces. Astonishingly, Moses not only visited the arms non-randomly, he didn't visit the same arm twice, implying that he had learned which arms still contained strawberry treats. In a second test in which potential visual landmarks were removed, the tortoise adopted a much simpler strategy, systematically moving from one arm to the next all the way around the apparatus. Wilkinson and colleagues expanded the study with more tortoises and found similar results.

Subsequent research has found that tortoises can follow gaze direction to learn from the behaviour of other animals. And in a paper called, 'The underestimated giants', researchers found evidence that giant tortoises not only remembered learned behaviours for up to nine years, they also learned faster when in a group, demonstrating an ability for social learning. Attitudes towards reptile research are changing too, although Wilkinson tells me that some researchers are still surprised by the idea of studying cognition in reptiles. 'People are generally intrigued, they find it quite funny, but when you show them the data they normally get really excited.'

Wilkinson's 28 red-footed tortoises are tested in a steamy 29°C lab, in contrast with the studies conducted in the mid-twentieth century, in which reptiles were tested in much colder facilities (it's therefore not surprising that they didn't perform well). As well as better integration of their natural ecology and behaviour into experiments, there's more of a focus on what we can learn from reptiles about the evolution of cognition. Tortoises, for example, have changed little in the last 200 million years, and while this was previously interpreted as evidence for their status as primitive animals that lack intelligence, another way to think about it is that tortoises may reveal the most ancient solutions to cognitive problems. For Wilkinson and others studying reptiles, this makes them extremely valuable animals for cognitive research.

What of the hare – how much truth is there to the expression 'hare-brained'? The difficulty with answering this, as for so many other animals, is that hare cognition is essentially unstudied. As Caravaggi says: 'People tend to be really interested in what the carnivores or omnivores can do. For herbivores, people tend to think "they're a bit thick, they're food, they all act the same" and it's not true. But nobody ever tries to show that it's not true.' Prey animals, in other words, tend to get the short shrift. This means that when it comes to thinking about how the hare might choose to behave, all we can do is speculate – the truth is that we just don't know.

★ ★ ★

Some have suggested that Aesop's fable is a metaphor for life. And when it comes to longevity, the tortoise certainly wins. The average lifespan of a hare is between four and seven years, depending on the species. Most species of tortoise, on the other hand, live between 80–150 years. What that means is that a hare could live its entire life before a red-footed tortoise has even evolved any sex characteristics. For the Galapagos giant tortoises, the largest species in the world, sex determination is not possible until at least 15 years, and the animal does not reach full adult size until it is 40 years old!

The giant tortoises are the longest-lived of the group,[*] comprising species on the Pacific islands and Galapagos. Rumour has it that Adwaita, an Aldabra giant tortoise, lived for an astonishing 255 years; however, this was never officially verified. A female radiated tortoise named Tu'i Malila, on the other hand, hatched in her native Madagascar in 1777 and was presented to the Tongan royal family by the British explorer Captain Cook. She remained in the care of the Tongan royal family until her death in 1965, aged 188. Then there's Jonathan, a Seychelles giant tortoise who is thought to have been born in 1832, making him the oldest known living tortoise. He lives with the governor of Saint Helena and three other tortoises, and is tended to by his own vet, who increasingly needs to hand-feed the venerable old beast. He's one of only a handful of individuals remaining, a sobering trend that is replicated across the world for the largest species. The other group of giants is found on the Galapagos and they're among the most iconic of this archipelago's wildlife. These islands were actually named for the tortoises: the old Spanish word *galapago* means saddle, a term used by early explorers for the tortoises due to the shape of their shells. They're thought to have drifted on debris to the Galapagos in the region of 2–3 million years ago, colonising different islands and subsequently diversifying into 15 separate species. Their longevity is demonstrated by the fact that Charles Darwin and Steve Irwin both cared for the same tortoise in their lives: Harriet was collected by Darwin in 1835 and ended up at Australia Zoo, which was founded by Irwin's parents. She died in 2006, aged at least 176 years (we don't

---

[*] Gigantism is one aspect of the general island rule proposed by J. Bristol Foster in 1964, which states that when mainland animals become isolated on islands, small species evolve to be larger than their mainland relatives, while large species evolve to be smaller (dwarfism).

know how old she was when Darwin collected her). Today, only 10 species of Galapagos tortoises are known for sure to still be in existence, all of which are under threat.* It's a dismal thought that these iconic gentle giants may one day be lost from the archipelago that bears their name.

The longevity of tortoises is a feature that has captivated people for centuries and influenced how the animals have been represented in mythology and folklore. Turtles are a symbol of the earth in many different Native American cultures, associated with long life, protection and fertility. In some Plains tribes, a newborn girl's umbilical cord was sewn into the shape of a turtle to ensure her health and safety. In other tribes, turtles are associated with healing, virtue, wisdom and spirituality.

White, on the other hand, despaired at Timothy's apparent squandering of such a long life on inactivity, writing that:

> It is a matter of wonder to find that Providence should bestow such a profusion of days, such a seeming waste of longevity, on a reptile that appears to relish it so little as to squander away more than two thirds of its existence in a joyless stupor, and be lost to all sensation for months together in the profoundest of slumbers.

White was approaching it in a purely anthropocentric fashion, of course. He was thinking of all the things he would do and achieve if he lived for 100 years or more. What he wasn't considering were the very distinct physiologies that restrict tortoises and other reptiles to living this way. On the great motorway of life, tortoises sit firmly in the crawler

---

*According to the latest report of the IUCN/SSC Tortoise and Freshwater Turtle Specialist Group, the turtle group is among the most threatened of the major vertebrate groups, with 59 per cent of the 356 living species classed as threatened.

lane, but their 'engine' is so efficient that they rarely need to stop and refuel. Hares can put their foot down and zoom past, but they don't get many miles to the gallon, meaning they must frequently pull in, fuel up and rest. We can't ever know what it would be like to live as a tortoise or, for that matter, a hare; all we can say about these contrasting ways of life is that each one works. The fact that the tortoise has barely changed in the last 200 million years should give us a clue as to the worth of its strategy. Perhaps we are the ones being a little bit 'hare' by bemoaning what we perceive as a wasted life: it seems very likely that on a geological timescale, the tortoise's footprint on this planet will outlive us all.

Fact or fiction?

No analysis of Aesop's animals would be complete without 'The Hare and the Tortoise'. Evaluating the truth to the fable is, however, not altogether straightforward. The tortoise is not designed for sprinting, while the hare is honed for it, and the differences go much further, covering evolutionary history, physiology, anatomy and behaviour. In most cases, I'm sure the hare would win; however, some of those differences might swing the race. It's not completely ludicrous to think that the hare could either hunker down mid-race or sprint off in the wrong direction, enabling the tortoise to just keep going to victory.

It can be difficult to keep up with the many versions of Aesop's fables created over the centuries. Through translation, elaboration and cultural influences the fables have shifted, acquired new subtleties, new meanings. You wouldn't think there's much that could be adapted for the tortoise and the hare, but in a version penned by the Irish writer Lord Dunsany in 1915, we get a glimpse of what may have occurred following the race.

In this version, the woodland animals support the tortoise, thinking he will win because of his hard shell. Just as in the original fable, the hare decides that he has time to take a nap, because the race is ridiculous. The tortoise wins and is celebrated by all as the fastest animal in the forest. Dunsany lets us know why we don't usually hear this 'real' version of the story, though:

> ... very few of those that witnessed it survived the great forest-fire that happened shortly after. It came up over the field by night with a great wind. The Hare and the Tortoise and a very few of the beasts saw it far off from a high bare hill that was at the edge of the trees, and they hurriedly called a meeting to decide what messenger they should send to warn the beasts in the forest.
>
> They sent the Tortoise.

Epilogue

It's before dawn in early January. I've opened the skylight in my bedroom and I'm peering down on a deserted street. The streetlights cast their glow through the misty drizzle; cloud cover is obscuring any other light. A moment later, I hear it. 'Wup wup wup!' The triple bark of a dog fox. He comes into view, weaving across the pavement and road, stopping to drink from a puddle before hurrying on his way. 'Wup wup wup!' as he disappears down the street. It's the middle of fox-breeding season and he's looking for a mate. I've not yet heard the wailing screech of a vixen, but the last two nights I've woken to hear him at around the same time – I appreciate his routine.

Watching this little animal, I can't help but admire its tenacity. I'm struck by Scott's plea, that we stop deriding the fox for being a crafty villain and instead consider its remarkable ability to thrive in our world. And it's not just the fox – time and time again the experts shared with me their hope that the animals they study be thought of in biological, not fictional, terms. After centuries of typecast characterisation, will it ever be possible for us to decouple our ingrained beliefs? To keep the characters of fables and folklore where they belong, as fictional creations in stories, and consider the real versions of these animals? Indeed, they are by no means less exciting.

I don't want to over-romanticise these animals and I'm certainly not advocating a switch to universal reverence. Wolves, foxes, lions and more cause considerable damage and have impacted the livelihoods of plenty of people throughout history. And as we consume more, encroaching further on their habitats and resources, human–wildlife conflict will increase. But myths and fairytales will not prevent livestock being taken; lamenting an animal's 'evil nature' will not help. Science, however, just might. By introducing data obtained from behavioural studies into conversations about wildlife management, we can make better-informed decisions and implement crucial non-lethal prevention and control measures. Understanding how different animals perceive and process information is our best chance of learning to communicate and live peacefully with them.

These are some hopes that keep me writing. My other ambition is simply that I have passed on some of my fascination with animal behaviour, and that you might go away from this book relaying a fact that resonated with you to someone else. Maybe the next time a video of a 'clever' critter pops up on your timeline you'll look at it differently and ask yourself what's going on behind its cute or funny behaviours.

Ducking my head out of the drizzle, I find myself wondering again what Aesop would think about all this – would he be surprised to learn the truth about his animals? Would he have chosen different characters if the science existed then? And how would more scientifically accurate portrayals have altered our world view? Stories are essential, powerful tools to entertain, inspire and teach, but to improve our understanding of our world, perhaps it's time that we melded the facts with the fables.

Selected Bibliography

Preface

Lefkowitz, J. B. 2018 'Reflection: listening to Aesop's animals'. In: *Animals: A History*, Adamson, P. & Edwards, G. F. (eds). Oxford University Press, New York.

Townsend, G. F. 1868. *Aesop's Fables*. George Routledge & Sons, London.

Chapter 1: The Crow and the Pitcher

The Avian Brain Nomenclature Consortium. 2005. Avian brains and a new understanding of vertebrate brain evolution. *Nature Reviews Neuroscience* 6: 151–159.

Bird, C. D & Emery, N. J. 2009. Rooks use stones to raise the water level to reach a floating worm. *Current Biology* 19: 1410–1414.

Cheke, L. G., Bird, C. D. & Clayton, N. S. 2011. Tool-use and instrumental learning in the Eurasian jay (*Garrulus glandarius*). *Animal Cognition* 14: 441–455.

Emery, N. 2016. Bird Brain: An Exploration of Avian Intelligence. Princeton University Press, New Jersey.

Heinrich, B. 1999. *Mind of the Raven: Investigations and adventures with wolf-birds*. HarperCollins, New York.

Hunt, G. R, 1996. Manufacture and use of hook-tools by New Caledonian crows. *Nature* 379: 249–251.

Marzluff, J. M. et al. 2012. Brain imaging reveals neuronal circuitry underlying the crow's perception of human faces. *Proceedings of the National Academy of Sciences of the United States of America* 109: 15912–15917.

Olkowicz, S. et al. 2016. Birds have primate-like numbers of neurons in the forebrain. *Proceedings of the National Academy of Sciences of the United States of America* 113: 7255–7260.

Taylor, A. H. et al. New Caledonian crows learn the functional properties of novel tool types. *PLoS ONE* 6: e26887.

Von Bayern, A. M. P. et al. 2018. Compound tool construction by New Caledonian crows. *Scientific Reports* 8: 15676

Weir, A. A. S., Chappell J. & Kacelnik A. 2002. Shaping of hooks in New Caledonian crows. *Science* 297: 981.

Chapter 2: The Wolf in Sheep's Clothing

Bugnyar, T., Reber, S.A. & Buckner, C. 2016. Ravens attribute visual access to unseen competitors. *Nature Communications* 7: 10506.

Byrne, R. W. & Whiten, A. (eds). 1988. *Machiavellian Intelligence: Social expertise and the evolution of intellect in monkeys, apes and humans.* Oxford University Press, Oxford.

Kano, F. et al. 2019. Great apes use self-experience to anticipate an agent's action in a false-belief test. *Proceedings of the National Academy of Sciences* 116: 20904–20909.

Krupenye, C. et al. 2016. Great apes anticipate that other individuals will act according to false beliefs. *Science* 354: 110–114.

Marshall-Pescini S. et al. 2017. Integrating social ecology in explanations of wolf–dog behavioral differences. *Current Opinion in Behavioral Sciences* 16: 80–86.

Mech, L. D. 1970. *The Wolf: The ecology and behavior of an endangered species.* Doubleday Publishing Co, New York.

Mech, L. D. 1999. Alpha status, dominance, and division of labor in wolf packs. *Canadian Journal of Zoology* 77: 1196–1203.

Premack, D. & Woodruff, G. 1978. Does the chimpanzee have a theory of mind? *Behavioral and Brain Sciences* 1: 515–526.

Range, F. et al. 2019. Wolves lead and dogs follow, but they both cooperate with humans. *Scientific Reports* 9: 3796.

Whiten, A. & Byrne, R. W. 1988. Tactical deception in primates. *Behavioral and Brain Sciences* 11: 233–273.

Chapter 3: The Dog and its Shadow

Bekoff, M. 2001. Observations of scent-marking and discriminating self from others by a domestic dog (*Canis familiaris*): tales of displaced yellow snow. *Behavioural Processes* 55: 75–79.

Berns, G. S., Brooks, A. M. & Spivak, M. 2015. Scent of the familiar: An fMRI study of canine brain responses to familiar and unfamiliar human and dog odors. *Behavioural Processes* 110: 37–46.

Gallup Jr, G. G. 1970. Chimpanzees: Self-recognition. *Science* 167: 86–87.

Gallup Jr, G. G. & Anderson, J. R. 2018. The 'olfactory mirror' and other recent attempts to demonstrate self-recognition in non-primate species. *Behavioural Processes* 148: 16–19.

Horowitz, A. 2017. Smelling themselves: Dogs investigate their own odours longer when modified in an 'olfactory mirror' test. *Behavioural Processes* 143: 17–24.

Horowitz, A. 2012. *Inside of a Dog*. Simon & Schuster, London.

Lea, S. E. G. & Osthaus, B. 2018. In what sense are dogs special? Canine cognition in comparative context. *Learning & Behavior* 46: 335–363.

Miklósi, Á. et al. 1998. Use of experimenter-given cues in dogs. *Animal Cognition* 1: 113–121.

Chapter 4: The Ass Carrying the Image

Bough, J. 2011. *Donkey*. Reaktion Books Ltd, London.

Brubaker, L. & Udell, M. A. R. 2016. Cognition and learning in horses (*Equus caballus*): What we know and why we should ask more. *Behavioural Processes* 126: 121–131.

Burden, F. & Thiemann, A. 2015. Donkeys are different. *Journal of Equine Veterinary Science* 35: 376–382.

Mejdell C. M. et al. Horses can learn to use symbols to communicate their preferences. *Applied Animal Behaviour Science* 184: 66–73.

Osthaus, B. et al. 2013. Spatial cognition and perseveration by horses, donkeys and mules in a simple A-not-B detour task. *Animal Cognition* 16: 301–303.

Proops, L. Burden, F. & Osthaus, B. Mule cognition: a case of hybrid vigour? *Animal Cognition* 12: 75–84.

Proops, L. et al. 2018. Animals remember previous facial expressions that specific humans have exhibited. *Current Biology* 28:1428–1432.

Rossel, S et al. 2008. Domestication of the donkey: Timing, processes, and indicators. *Proceedings of the National Academy Sciences U S A*. 105: 3715–3720.

Mitchell, P. 2018. *Donkey in Human History: An archaeological perspective*. Oxford University Press, Oxford and New York.

Chapter 5: The Fox and the Crow

Baker, P. J., Newman, T., & Harris, S. 2001. Bristol's foxes – 40 years of change. *British Wildlife* 12: 411–417.

Červený, J. et al. 2011. Directional preference may enhance hunting accuracy in foraging foxes. *Biology Letters* 7: 355–357.

Dugatkin, L. A. & Trut, L. 2017. *How to tame a fox (and build a dog): visionary scientists and a Siberian tale of jump-started evolution.* University of Chicago Press, Chicago.

Hare, B. et al. 2005. Social cognitive evolution in captive foxes is a correlated by-product of experimental domestication. *Current Biology* 15: 226–230

Henry, J. D. 1996. *Red Fox: The catlike canine.* Smithsonian Books, Washington DC.

Macdonald, D. 1987. *Running with the Fox.* First Edition. Unwin Hyman, London.

Parsons, K. J. et al. 2020. Skull morphology diverges between urban and rural populations of red foxes mirroring patterns of domestication and macroevolution. *Proceedings of the Royal Society B: Biological Sciences* 287: 20200763.

Scott, D. M. et al. 2018. A citizen science–based survey method for estimating the density of urban carnivores. *PLoS ONE* 13: e0197445.

Chapter 6: The Lion and the Shepherd

Borrego, N. & Gaines, M. 2016. Social carnivores outperform asocial carnivores on an innovative problem. *Animal Behaviour* 114: 21–26.

Brosnan, S. F., de Waal, F. B. M. 2003. Monkeys reject unequal pay. *Nature* 425: 297–299.

Carter, G. & Wilkinson, G. 2013. Does food sharing in vampire bats demonstrate reciprocity? *Communicative & Integrative Biology* 6: e25783.

Grinnell, J., Packer, C. & Pusey, A. E. 1995. Cooperation in male lions: kinship, reciprocity or mutualism? *Animal Behaviour* 49: 95–105.

Packer, C. 1994. *Into Africa.* University of Chicago Press, Chicago

Packer, C. 2010. Lions. *Current Biology* 20: R590–R591.

Schaller, G. 1972. *The Serengeti Lion*. University of Chicago Press, Chicago.

Schweinfurth, M. K. & Call, J. 2019. Reciprocity: Different behavioural strategies, cognitive mechanisms and psychological processes. *Learning & Behavior* 47: 284–301.

Trivers, R. L. 1971. The Evolution of Reciprocal Altruism. *The Quarterly Review of Biology* 46: 35–57.

Wilkinson, G. S. 1984. Reciprocal food sharing in the vampire bat. *Nature* 308: 181–184.

Chapter 7: The Monkey and the Fisherman

Fragaszy, D., Visalberghi, E. 2004. Socially biased learning in monkeys. *Learning & Behavior* 32: 24–35.

Galef, B. G. 1988. 'Imitation in animals: history, definition, and interpretation of data from the psychological laboratory'. Zentall, T. R. & Galef, B. G. (eds) *Social Learning: Psychological and Biological Perspectives*. Hillsdale, New Jersey.

Haggerty, M. E. 1912. Imitation and animal behavior. *Journal of Philosophy* 9: 265–272.

Hirata, S., Watanabe, K. & Masao, K. 2008. 'Sweet-Potato Washing' Revisited. In: Matsuzawa T. (eds) *Primate Origins of Human Cognition and Behavior*. Springer, Tokyo.

Perry, S. et al. 2003. Social conventions in wild white-faced capuchin monkeys. *Current Anthropology* 44: 241–268.

Van de Waal, E., Borgeaud, C. & Whiten, A. 2013. Potent social learning and conformity shape a wild primate's foraging decisions. *Science* 340: 483–485.

Van de Waal, E., Claidière, N. & Whiten, A. 2015. Wild vervet monkeys copy alternative methods for opening an artificial fruit. *Animal Cognition* 18: 617–627.

Whiten, A. et al. 1999. Cultures in chimpanzees. *Nature* 399: 682–685.

Chapter 8: The Ants and the Grasshopper

Clayton, N. S., Bussey, T. J. & Dickinson, A. 2003. Can animals recall the past and plan for the future? *Nature Reviews Neuroscience* 4: 685–691.

Franks, N. R. et al. 2007. Reconnaissance and latent learning in ants. *Proceedings of the Royal Society B: Biological Sciences* 274: 1505–1509.

Kabadayi, C. & Osvath, M. 2017. Ravens parallel great apes in flexible planning for tool-use and bartering. *Science* 357: 202–204.

Mulcahy, N. J. & Call, J. 2006. Apes save tools for future use. *Science* 312: 1038–1040.

Osvath, M. Spontaneous planning for future stone throwing by a male chimpanzee. *Current Biology* 19: R190–R191.

Roberts, W. A. 2012. Evidence for future cognition in animals. *Learning and Motivation* 43: 169–180.

Chapter 9: The Hare and the Tortoise

Bejan, A. et al. 2018. The fastest animals and vehicles are neither the biggest nor the fastest over lifetime. *Scientific Reports* 8: 12925.

Caravaggi, A. 2018. Lagomorpha Life History. In: Vonk J. & Shackelford T. (eds) *Encyclopedia of Animal Cognition and Behavior*. Springer, Cham, Switzerland.

Caravaggi, A. 2018. Lagomorpha Navigation. In: Vonk J. & Shackelford T. (eds) *Encyclopedia of Animal Cognition and Behavior*. Springer, Cham, Switzerland.

Matsubara, S., Deeming, D. C. & Wilkinson, A. Cold-blooded cognition: new directions in reptile cognition. *Current Opinion in Behavioral Sciences* 16: 126–130.

Taylor, M. 2017. *The Way of the Hare*. Bloomsbury, London.

Acknowledgements

This book has been a long time coming, a work in progress that was conceived in 2012. I remember chatting to my dad about the idea while we were out walking one misty, autumnal Herefordshire day but other things were going on, so I was only able to tease it in my mind and start, gently, pulling thoughts together. When my dad died in 2016 my mental health took a nosedive. Writing a book seemed like a lost cause but I wanted my idea to be out there, so I pitched it as an essay to *BBC Wildlife*. I must thank Ben Hoare for taking a chance on an unknown writer: seeing the essay in print gave me the confidence boost I needed at the time I needed it most.

Ben recommended that I connect with Jim Martin, who had started a popular science imprint at Bloomsbury. It is not as neat a story as that but when, early in 2018, Jim tweeted that he was seeking new popular science submissions, I finally got my act together. Thank you, Jim – your enthusiasm for the idea and flexibility through this has been hugely appreciated – and everyone in the Bloomsbury team. Special thanks to my editors, initially Anna MacDiarmid and latterly Angelique Neumann, for everything they did to bring this to print. I couldn't have asked for a better copyeditor in Emily Kearns, whose scrutiny of the manuscript was invaluable, and I'm also thankful for Jo Mortimer's expert proof-reading. On the publicity and marketing side, Amy Greaves and Alice Graham have been fantastically helpful and patient guides. Finally, the book looks fantastic, and for that the whole design and production team deserves thanks. Each chapter has also been beautifully brought to life by Hana Ayoob's illustrations – I couldn't be happier that she agreed to work on this.

Over the past few years, I have contacted many experts and, fortunately, most of them responded warmly. So my heartfelt thanks go to everyone who took the time to speak with me or email responses to my questions: Marc Bekoff, Greg Berns,

Chris Bird, Natalia Borrego, Thomas Bugnyar, Faith Burden, Dick Byrne, Anthony Caravaggi, Gerry Carter, Lucy Cheke, Gordon Gallup, Ben Hart, Alexandra Horowitz, Alex Kacelnik, Chris Krupenye, John Marzluff, Dave Mech, Mathias Osvath, Craig Packer, Loma Pendergraft, Leanne Proops, Friederike Range, Manon Schweinfurth, Dawn Scott, Kaeli Swift, Erica van de Waal, Elisabetta Visalberghi, Andy Whiten, Anna Wilkinson. Thank you all for your generosity of time and spirit – the book is so much better for your contributions. Several of you also took the time to read through and sense-check chapter drafts or sections, for which I am especially appreciative. I'd like to single out Dick Byrne and Alex Kacelnik for special thanks – you both went above and beyond in your willingness to help, and your detailed reviews and constructive critiques were instrumental in helping me to crystallise my thoughts.

Thanks are also due to everyone in the writing and publishing industry who took the time to provide advice and support. JV Chamary, my mentor in the ABSW pilot scheme, has been a fantastic guide, offering the perfect balance of honest feedback and encouragement. And I'm indebted to fellow Sigma author Jules Howard, who took time away from his own looming deadline to review some of my work.

I've been lucky to have a wonderful support network of family and friends, some of whom enthusiastically also signed up to read chapter drafts – Mum, Josh and Annie, thank you so much for your encouragement and candid feedback, it's exactly what I needed. Outside of the book, regular family video calls have provided welcome respite, while friends have been on hand for calls, walks, coffee and beer. Thank you to those that tolerated me imposing strict time limits on our strolls and who understood my increased flakiness as my deadline approached. I'm grateful that every one of you is in my life, knowing when to ask about the book and when to coax me out of my book-shaped bubble.

Finally, to John, who has been there every step of the way. When I got stuck, you lent an ear and an alternative view. When I was flagging, you gave me hugs. When I was done, you poured the champagne. You are and always will be my rock.

Index